INTRODUCTION TO FIELD THEORY

INTRODUCTION TO FIELD THEORY

IAIN T. ADAMSON

Senior Lecturer in Mathematics, University of Dundee

SECOND EDITION

CAMBRIDGE UNIVERSITY PRESS

Cambridge

London New York New Rochelle

Melbourne Sydney

CAMBRIDGE UNIVERSITY PRESS
Cambridge, New York, Melbourne, Madrid, Cape Town, Singapore, São Paulo, Delhi

Cambridge University Press
The Edinburgh Building, Cambridge CB2 8RU, UK

Published in the United States of America by Cambridge University Press, New York

www.cambridge.org
Information on this title: www.cambridge.org/9780521286589

First published by Oliver & Boyd 1964
Second edition published by Cambridge University Press 1982
Re-issued in this digitally printed version 2008

A catalogue record for this publication is available from the British Library

Library of Congress Catalogue Card Number: 82–1164

ISBN 978-0-521-24388-9 hardback (first edition)
ISBN 978-0-521-28658-9 paperback (second edition)

CONTENTS

CHAPTER IV

APPLICATIONS

PREFACE

AMID all the current interest in modern algebra, field theory has been rather neglected—most of the recent textbooks in algebra have been concerned with groups or vector spaces. But field theory is a very attractive branch of algebra, with many fascinating applications; and its central result, the Fundamental Theorem of Galois Theory, is by any standards one of the really "big" theorems of mathematics. This book aims to bring the reader from the basic definitions to important results and to introduce him to the spirit and some of the techniques of abstract algebra. It presupposes only a little knowledge of elementary group theory and a willingness on the reader's part to remember definitions precisely and to engage in close argument.

Chapter 1 develops *ab initio* the elementary properties of rings, fields and vector spaces. Chapter 2 describes extensions of fields and various ways of classifying them. In Chapter 3 we give an exposition of the Galois theory of normal separable extensions of finite degree, closely following Artin's approach. Chapter 4 provides a wide variety of applications of the preceding theory, including the classification of all fields with a finite number of elements, ruler-and-compasses constructions and the impossibility of solving by radicals the generic polynomial of degree greater than 4.

I had the great good fortune to persuade my colleague Dr Hamish Anderson to read the first draft of this book, and as a result of his careful scrutiny and penetrating comments many blemishes were removed; I am deeply grateful to him for his invaluable help. In preparing the book for the press I have had the assistance of Dr Joan Aldous, Mr William Blackburn and Mr Brian Kennedy, and I gratefully acknowledge this here. It is a pleasure also to record my great gratitude to Professor D. E. Rutherford, whose lectures on groups first aroused my interest in algebra, for his constant encouragement and help at all stages in the preparation of the book. Finally I must not forget to thank several generations of honours students in The Queen's University of Belfast and Queen's College, Dundee who patiently listened to and commented on the successive lecture courses which eventually turned into this book; one of them in particular, in a spontaneous exclamation, provided me with an appropriate conclusion to Chapter 4.

IAIN T. ADAMSON

DUNDEE
August 1964

CHAPTER I

ELEMENTARY DEFINITIONS

§ **1. Rings and fields.** Modern algebra, of which field theory is a part, may be very roughly described as the study of sets equipped with laws of composition. To amplify this description we make the following definition: a **law of composition** on a set E is an operation which assigns to every ordered pair (a, b) of elements of E a definite element of E which may be denoted by $a+b$, in which case it is called the **sum** of a and b and the operation is called **addition**, or alternatively by $a \times b$ or $a \,.\, b$ or simply ab, in which case it is called the **product** of a and b and the operation is called **multiplication.** Quite clearly ordinary addition and multiplication of real numbers are laws of composition on the set of real numbers $\mathbf{R}$.

When we wish to discuss laws of composition in general, we use a " neutral " symbol, such as $a \circ b$, to denote the result of applying the law of composition to the ordered pair (a, b). With this notation we make some further definitions. A law of composition on a set E is said to be **associative** if for every three elements a, b, c of E we have $a \circ (b \circ c) = (a \circ b) \circ c$; it is said to be **commutative** if for every pair of elements a, b of E we have $a \circ b = b \circ a$. If for any two elements c, d of E we have $c \circ d = d \circ c$ then c and d are said to **commute**. An element n of E is called a **neutral element** for the law of composition if $n \circ a = a = a \circ n$ for every element a of E; if the additive notation is used, a neutral element is called a **zero element** and is usually denoted by 0, and if the multiplicative notation is used, a neutral element is called an **identity element** and is denoted by e or 1. If a is an element of E, an **inverse** of a relative to a law of composition for which there is a neutral element n is an element a' of E such that $a \circ a' = n = a' \circ a$; when the additive or multiplicative notations are used we write $-a$ or a^{-1} respectively instead of a'. Ordinary addition and multiplication of real numbers are both associative and commutative; the real numbers 0 and 1 are neutral elements for addition and multiplication respectively; every real number has an inverse relative to addition and every real number except 0 has an inverse relative to multiplication. Addition and multiplication of real numbers have a further property: for every three real numbers a, b, c we have

$$a(b+c) = ab+ac \text{ and } (b+c)a = ba+ca.$$

We say that the multiplication is **distributive** with respect to the addition.

Readers of this book will require to have some familiarity with the elementary theory of groups, as contained in practically any introductory text in modern algebra. We recall here that a **group** is a set G equipped with an associative law of composition such that

(1) there is a neutral element for the law of composition;

(2) every element has an inverse relative to the law of composition.

It is unnecessary to state explicitly the "closure" property mentioned by some writers on elementary group theory, since it is built into our definition of a law of composition on G that the result of applying the law to a pair of elements of G is again an element of G. If the law of composition of a group is commutative, the group is said to be **abelian**.

A **ring** is a set R equipped with two laws of composition, which we shall call addition and multiplication, such that the following conditions are satisfied:

A1. The addition is associative, i.e., for every three elements a, b, c of R we have $a+(b+c) = (a+b)+c$.

A2. The addition is commutative, i.e., for every pair of elements a, b of R we have $a+b = b+a$.

A3. There is a neutral element for the addition, i.e., an element, which we call zero and denote by 0, such that for every element a of R we have $a+0 = a = 0+a$.

A4. Every element a of R has an inverse relative to the addition, i.e., an element which we denote by $-a$ such that $a+(-a) = 0 = (-a)+a$.

M1. The multiplication is associative, i.e., for every three elements a, b, c of R we have $a(bc) = (ab)c$.

AM. The multiplication is distributive with respect to the addition, i.e., for every three elements a, b, c of R we have $a(b+c) = ab+ac$ and $(b+c)a = ba+ca$.

A ring R is called a **commutative ring** if, in addition to the defining properties of a ring, it satisfies the further condition:

M2. The multiplication is commutative, i.e., for every pair of elements a, b of R we have $ab = ba$.

A ring R is called a **ring with identity** if it satisfies the conditions **A1**, **A2**, **A3**, **A4**, **M1**, **AM** and the further condition:

M3. There is a neutral element for the multiplication, i.e., an element e, which we call the identity of R, such that for every element a of R we have $ea = a = ae$.

Finally, a commutative ring with identity is called a

field if it contains at least two elements and satisfies all the conditions listed so far, together with the following:

M4. Every non-zero element a of R has an inverse relative to the multiplication, i.e., an element a^{-1} such that $aa^{-1} = e = a^{-1}a$.

A field F is said to be **finite** or **infinite** according as the number of elements of F is finite or infinite.

Example 1. The most familiar example of a ring is the set of ordinary integers (positive, negative and zero) equipped with their ordinary operations of addition and multiplication; we shall always denote this ring by $\mathbf{Z}$. It is a commutative ring with identity (the number 1 is clearly the identity), but not a field—indeed the only integers with multiplicative inverses in $\mathbf{Z}$ are 1 itself and -1.

Example 2. The sets of rational numbers, real numbers and complex numbers, with the ordinary definitions of addition and multiplication, are easily seen to satisfy all the conditions for fields. We denote these fields respectively by $\mathbf{Q}$, $\mathbf{R}$ and $\mathbf{C}$.

Example 3. If R is any ring and n is any positive integer, then the set of $n \times n$ matrices with elements in R, equipped with ordinary matrix addition and multiplication, is a ring which we denote by $M_n(R)$. The ring $M_n(R)$ is in general not commutative; it has an identity if R does.

Example 4. Let m be any positive integer greater than 1; if a and b are integers such that $a-b$ is divisible by m we say that a is **congruent** to b **modulo** m, and we write $a \equiv b$ mod. m. The **residue class** of an integer a modulo m is the set of all integers congruent to a modulo m; it is clear that there are exactly m distinct residue classes, since every integer is congruent modulo m to precisely one of the integers $0, 1, \ldots, m-1$. We denote the set of residue classes modulo m by $\mathbf{Z}_m$ and we proceed to turn it into a ring by defining appropriate operations of addition and multiplication. Let C_1 and C_2 be any two residue classes; choose any integer a_1 from C_1 and any integer a_2 from C_2; we define C_1+C_2 and C_1C_2 to be the residue classes of a_1+a_2 and a_1a_2 respectively. At first sight it appears that these residue classes may depend upon the choice of a_1 and a_2, but we shall show that this is not in fact so. Namely, if b_1, b_2 are integers in the residue classes C_1, C_2 respectively, then $a_1 \equiv b_1$ and $a_2 \equiv b_2$ mod. m, and so there are integers k_1 and k_2 such that $a_1 = b_1+k_1m$ and $a_2 = b_2+k_2m$. It follows that $a_1+a_2 = b_1+b_2+(k_1+k_2)m$ and $a_1a_2 = b_1b_2+(k_1b_2+k_2b_1+k_1k_2m)m$; so we have $a_1+a_2 \equiv b_1+b_2$ mod. m and $a_1a_2 \equiv b_1b_2$ mod. m. Thus a_1+a_2 and b_1+b_2 belong to the same residue class modulo m and so do a_1a_2 and b_1b_2. Hence C_1+C_2 and C_1C_2 depend only upon C_1 and C_2 and not upon the choices involved in their definition.

It is easy to verify that with these laws of composition $\mathbf{Z}_m$ is a commutative ring with identity; the zero and identity elements O and E are the residue classes containing the integers 0 and 1 respectively, and the additive inverse of the residue class containing a is the residue class containing $-a$.

We say that an integer is **relatively prime** to m if it has no factor in common with m except 1 and -1. It is clear that if one integer in a residue class C modulo m is relatively prime to m then so are all the integers in C; in this case we say that C is a **relatively prime residue class.** Let $\mathbf{R}_m$ be the set of relatively prime residue classes modulo m. We shall now show that a residue class modulo m has a multiplicative inverse in $\mathbf{Z}_m$ if and only if it belongs to $\mathbf{R}_m$.

Suppose first that the residue class C has a multiplicative inverse C' in $\mathbf{Z}_m$; then $CC' = E$, and so, if a and a' are integers in C and C' respectively, we have $aa' \equiv 1$ mod. m.

Thus there is an integer k such that $aa' + km = 1$. It follows that if r is any common factor of a and m then r is a factor of $aa'+km = 1$; hence r is either 1 or -1. So if C has a multiplicative inverse in $\mathbf{Z}_m$, C belongs to $\mathbf{R}_m$. We note incidentally that since C' also has an inverse in $\mathbf{Z}_m$ (C is the inverse of C'), C' also belongs to $\mathbf{R}_m$.

Conversely, suppose that C belongs to $\mathbf{R}_m$; we shall show that C has a multiplicative inverse in $\mathbf{Z}_m$. To this end we choose an integer a from C; then a is relatively prime to m. We now consider the set of positive integers which can be expressed in the form $xa+ym$ where x and y are integers. This set is clearly non-empty and hence contains a least element, $d = x_0a+y_0m$ say. Dividing a by d we obtain integers q, r such that $a = qd+r$, $0 \leqq r < d$; from this it follows that

$$r = a-q(x_0a+y_0m) = (1-qx_0)a+(-qy_0)m.$$

If r were non-zero this would contradict our choice of d as the least positive integer of the form $xa+ym$; hence $r = 0$ and d is a factor of a. An exactly similar argument shows that d is a factor of m. Thus d is a positive common factor of a and m and so $d = 1$. We have now shown that there exist integers x_0 and y_0 such that $x_0a+y_0m = 1$, whence $x_0a \equiv 1$ mod. m. Hence if C' is the residue class of x_0 modulo m we have $CC' = E$; that is to say, C' is a multiplicative inverse of C.

Since the product of two integers relatively prime to m is also relatively prime to m it follows that the product of two residue classes in $\mathbf{R}_m$ is also in $\mathbf{R}_m$; thus the multiplication in $\mathbf{Z}_m$ is an associative law of composition on $\mathbf{R}_m$. The identity residue class E belongs to $\mathbf{R}_m$ and the preceding arguments show that every residue class in $\mathbf{R}_m$ has a multiplicative inverse which is also in $\mathbf{R}_m$. It follows that $\mathbf{R}_m$, equipped with the multiplication operation of $\mathbf{Z}_m$, is an abelian group.

Example 5. Let p be a prime number and form the

ring $\mathbf{Z}_p$ according to the procedure described in Example 4; we claim that $\mathbf{Z}_p$ is a field. Since we have already proved that $\mathbf{Z}_p$ is a commutative ring with identity, the only property which remains to be established is **M4**. But this follows at once from our discussion in Example 4 since every non-zero residue class modulo p is relatively prime.

§ **2. Elementary properties.** We notice that conditions **A1** to **A4** can be summed up by saying that the set of elements of a ring R, equipped only with the addition operation, forms an abelian group. This group is called the **additive group** of the ring and is denoted by R^+.

It is easy to show that the zero element 0 of R and the additive inverse $-a$ of each element a of R are unique. Suppose first that 0 and $0'$ are zero elements. Then $0+0' = 0$ (since $0'$ is a zero element) and $0+0' = 0'$ (since 0 is a zero element); hence $0 = 0'$. Next suppose $-a$ and a' are additive inverses of a. Then we have $(a'+a)+(-a) = 0+(-a) = -a$ (since a' is an inverse) and $(a'+a)+(-a) = a'+(a+(-a)) = a'+0 = a'$ (since the addition is associative and $-a$ is an inverse); hence $a' = -a$. It now follows at once that for every element a of R we have $-(-a) = a$; for both these elements are additive inverses of $-a$.

The existence of an additive inverse for every element in a ring implies that in a ring subtraction is always possible. For the problem of subtracting an element b from an element a can be reformulated as the problem of finding an element x such that $a = x+b$; and clearly $x = a+(-b)$ satisfies this requirement, since

$$(a+(-b))+b = a+((-b)+b) = a+0 = a.$$

We usually abbreviate $a+(-b)$ to $a-b$. Since the addition operation in a ring is commutative, we have $a-b = (-b)+a$.

Although the zero element of a ring is originally singled out for special attention by virtue of its additive property, the distributive condition **AM** implies that it also enjoys the multiplicative property which we are accustomed to associate with the real number zero—namely, that if one of the factors in a product is zero then the product is zero. So let a be any element of R; we shall prove that $a0 = 0$. Since 0 is a neutral element for addition, we have $0+0 = 0$ and hence $a(0+0) = a0$. Using **AM** we deduce that $a0+a0 = a0$. Now, by **A4**, $a0$ has an additive inverse $-(a0)$; adding $-(a0)$ to both sides, we obtain

$$-(a0)+(a0+a0) = -(a0)+a0 = 0.$$

Applying the associative condition **A1** on the left side, we have $(-(a0)+a0)+a0 = 0$, whence $0+a0 = 0$, i.e., $a0 = 0$ as we claimed. Similarly we may show that $0a = 0$ for every element a of R. It follows that in a field the zero and identity elements 0 and e are distinct; for if a is a non-zero element we have $a0 = 0$, but $ae = a$.

Rather similar arguments can be used to prove that if a and b are any two elements of a ring R then $a(-b) = -(ab)$, $(-a)b = -(ab)$ and $(-a)(-b) = ab$. For example, to establish the first of these, we notice that $b+(-b) = 0$ and hence $a(b+(-b)) = a0$. Thus, by **AM** and what we have just proved, $ab+a(-b) = 0$; so $a(-b)$ is an additive inverse of ab. But $-(ab)$ is the unique additive inverse of ab; hence $a(-b) = -(ab)$. The other results are obtained by analogous arguments.

We have shown for every ring R that if one of the factors in a product of elements of R is zero then the product is zero. The converse of this result is not true in general; for example, in the ring $M_2(\mathbf{C})$ of 2×2 matrices with complex elements, we have

$$\begin{bmatrix} 1 & i \\ i & -1 \end{bmatrix}\begin{bmatrix} 1 & i \\ i & -1 \end{bmatrix} = \begin{bmatrix} 0 & 0 \\ 0 & 0 \end{bmatrix}.$$

The converse *is* true, however, in the case of fields. For, suppose a and b are elements of a field F such that $ab = 0$; we shall show that either $a = 0$ or $b = 0$—in other words, that if a is non-zero then $b = 0$. If a is non-zero, then it has a multiplicative inverse a^{-1}. Multiplying by a^{-1} we obtain $a^{-1}(ab) = a^{-1}0$, and hence, by **M1**,

$$b = eb = (a^{-1}a)b = a^{-1}(ab) = a^{-1}0 = 0.$$

This discussion shows that the product of two non-zero elements of a field is non-zero. Hence in a field the multiplication operation is a law of composition on the set of non-zero elements. Conditions **M1**, **M2**, **M3** and **M4** now imply that the set of non-zero elements of a field F, equipped only with the multiplication operation, forms an abelian group. We call this group the **multiplicative group** of the field F and denote it by F^*. By arguments similar to those used in the additive group we can easily establish that the identity element e and the multiplicative inverse of each non-zero element of a field F are unique, and that, for every non-zero element a of F, $(a^{-1})^{-1} = a$.

Finally, in a field F, division (except by the zero element) is always possible. To divide an element a by a non-zero element b we must find an element x such that $a = xb$; $x = ab^{-1}$ clearly satisfies this requirement. Since the multiplication operation in a field is commutative, we have $ab^{-1} = b^{-1}a$. We frequently use the " fraction " notation a/b instead of ab^{-1}.

Let F be any field. We now define an operation which assigns to each integer n and each element a of F an element of F which we denote by na. We make the definition inductively by setting

(i) $0a = 0$;

(ii) $(k+1)a = ka+a$ for all integers $k \geqq 0$;

(iii) $(-k)a = -(ka)$ for all integers $k > 0$.

We call the elements na the **integral multiples** of a. It can be shown by mathematical induction that for all integers m, n and all elements a, b of F we have

$$(m+n)a = ma+na,\ m(a+b) = ma+mb,$$
$$(mn)a = m(na),\ (ma)(nb) = (mn)(ab).$$

We now consider the integral multiples of the identity e of a field F. We examine in particular the set of integers n such that $ne = 0$; we denote this set by A and call it the **annihilator** of e. Two cases can arise.

Case 1. The annihilator consists of the integer 0 alone. In this case we say that the field F has **characteristic zero** and we note that if n is a non-zero integer and a is a non-zero element of F then $na = n(ea) = (ne)(1a)$ is non-zero.

Case 2. The annihilator contains at least one non-zero integer. It follows at once that it contains at least one positive integer; for, if a is in A, then $ae = 0$, whence $(-a)e = -(ae) = 0$, and so $-a$ is in A also; but, if a is non-zero, one of the integers a and $-a$ is positive. Let p be the least positive integer in A. We shall show that p is a prime number and that A consists precisely of the multiples of p.

First, suppose p is not a prime number, so that p can be expressed in the form $p = ab$ where $1<a,\ b<p$. Then we have $0 = pe = (ab)e = (ae)(be)$, whence either $ae = 0$ or $be = 0$. But this contradicts the choice of p as the least positive integer in A, and hence p is a prime number.

Next, it is clear that all the multiples of p belong to A since for every integer n we have $(np)e = n(pe) = n0 = 0$. Conversely, suppose k belongs to A. Then if q is the quotient and r the remainder when k is divided by p we have $k = qp+r$ and $0 \leq r<p$. Hence

$$0 = ke = (qp+r)e = (qp)e+re = re,$$

so that r belongs to A. If r were non-zero this would

again contradict the choice of p as the least positive integer in A; hence $r = 0$ and k is a multiple of p.

In this case we say that the field F has **characteristic** p, and we notice that if k is any multiple of p and a is any element of F then $ka = 0$.

Let S be a subset of a ring R. If a and b are any two elements of S their sum and product $a+b$ and ab are certainly elements of R because the addition and multiplication are laws of composition on R. It may happen, however, that for every pair of elements a, b of S the sum and product actually belong to S itself. In this case we can regard the addition and multiplication as laws of composition on the subset S, and it then makes sense to enquire whether S, equipped with these laws of composition, is a ring or even a field; if so, we say that S is a **subring** or **subfield** of R. We remark that in examining whether S is a subring or a subfield of R we need not concern ourselves with the "formal" conditions **A1**, **A2**, **M1** and **AM**—these will be satisfied automatically since the laws of composition on R already satisfy them; this is true also of **M2** if R is commutative.

To show that a subset F of a field K is a subfield it is thus sufficient to verify that

(1) for every pair of elements a, b of F, $a+b$ and ab belong to F;

(2) for every element a of F, $-a$ belongs to F;

(3) for every non-zero element a of F, a^{-1} belongs to F.

This is clear because (1) and (2) together imply that $a+(-a) = 0$ belongs to F and (1) and (3) similarly imply that $aa^{-1} = e$ belongs to F.

Clearly every field K has at least one subfield—namely K itself. A field which contains no subfield except itself is called a **prime field**.

Example 1. The field of rational numbers **Q** is a prime field. Suppose F is any subfield of **Q**. Let a be any non-zero element of F; since F is a subfield of **Q** it contains the inverse of a, and hence the product $aa^{-1} = 1$. It now follows that F contains all the integers, and hence the inverses of all the non-zero integers; but this implies that F contains every rational number $p/q = pq^{-1}$. Hence $F = \mathbf{Q}$ and so **Q** is a prime field.

Example 2. For every prime number p, the field $\mathbf{Z}_p$ is a prime field. It is easy to convince oneself that if F is a subfield of $\mathbf{Z}_p$ then the additive group F^+ of F is a subgroup of the additive group $\mathbf{Z}_p^+$ of $\mathbf{Z}_p$. Now $\mathbf{Z}_p^+$ is a group of prime order p and hence has no subgroups except itself and the subgroup consisting of the zero element alone. Hence $F = \mathbf{Z}_p$ and so $\mathbf{Z}_p$ is a prime field. (A subfield F cannot consist of the zero element alone, since, by definition, a field contains at least two elements.)

In the next section we shall show that in a certain sense there are no prime fields other than the ones we have just described.

We conclude the present section by proving that every field includes a prime field. Let K be any field and consider the subset P of K consisting of those elements which are common to all the subfields of K. If a and b are any elements of P then a and b belong to every subfield L of K, and so it follows that $a+b$, ab, $-a$ and a^{-1} (if a is non-zero) also belong to every subfield L and hence to P itself. This shows that P is a subfield of K; according to the definition of P every subfield of K includes P. From this we deduce that P is a prime field. For if F is any subfield of P it is also a subfield of K and hence includes P; since F both includes and is included in P it follows that $F = P$, and so P has no subfield except itself. We call P the **prime subfield** of K.

§ 3. **Homomorphisms.** Let A and B be any two sets. A **mapping** of A into B is an operation ϕ which assigns to each element a of A a well-determined element of B which we denote by $\phi(a)$ and call the **image** of a under ϕ. If A_1 is any subset of A the subset of B consisting of the images under ϕ of all the elements of A_1 is called the image of A_1 under ϕ and is denoted by $\phi(A_1)$; if $\phi(A) = B$ —in other words, if every element in B is the image under ϕ of at least one element of A—we say that ϕ maps A **onto** B. A mapping ϕ is said to be **one-to-one** or (1, 1) if distinct elements of A always have distinct images under ϕ; thus ϕ is one-to-one if, whenever we have $\phi(a_1) = \phi(a_2)$, we can conclude that $a_1 = a_2$. If ϕ is a mapping of A into B we sometimes write $\phi:A\to B$ or $A\xrightarrow{\phi}B$. Two mappings ϕ_1 and ϕ_2 of A into B are said to be **equal** if for each element a of A the images of a under ϕ_1 and ϕ_2 are equal; that is to say, $\phi_1 = \phi_2$ if and only if $\phi_1(a) = \phi_2(a)$ for every element a of A.

Now let R_1 and R_2 be rings and let ϕ be a mapping of R_1 into R_2. If we are given two elements a and b of R_1 two procedures naturally suggest themselves. First, we may form the sum and product of these elements in R_1 and then apply the mapping ϕ to the results; we obtain the elements $\phi(a+b)$ and $\phi(ab)$ in R_2. Alternatively, we may apply ϕ to a and b and then form the sum and product of the images $\phi(a)$ and $\phi(b)$ in R_2; we obtain $\phi(a)+\phi(b)$ and $\phi(a)\phi(b)$. For an arbitrary mapping ϕ there is no reason to suppose that these procedures will lead to the same destinations; mappings for which they do are called homomorphisms. That is to say, a **homomorphism** of a ring R_1 into a ring R_2 is a mapping ϕ of R_1 into R_2 such that, for every pair of elements a, b of R_1, we have

$$\phi(a+b) = \phi(a)+\phi(b) \text{ and } \phi(ab) = \phi(a)\phi(b).$$

A homomorphism which maps R_1 *onto* R_2 is called an

epimorphism; a homomorphism which is one-to-one is called a **monomorphism**; a homomorphism which is both an epimorphism and a monomorphism is called an **isomorphism**. If there is an isomorphism of a ring R_1 onto a ring R_2 we say that R_1 and R_2 are **isomorphic**. If $R_1 = R_2 = R$, say, an isomorphism of R onto R is called an **automorphism** of R.

Readers without a classical background may find the last paragraph a little overwhelming and may welcome a brief comment on the derivation of the terms introduced there. The " morphism " part common to them all is derived from the Greek μορφή (*morphe*), meaning " form " or " structure "; the prefixes " homo- ", " epi- ", " mono- ", " iso- " and " auto- " are English versions of Greek words meaning " similar ", " onto ", " single ", " equal " and " self " respectively. We notice in particular that, in a sense made quite precise above, isomorphic rings have " equal structure "; they differ only in the nature of their elements and for many purposes may be regarded as " essentially the same ".

The terminology introduced above is also used for mappings of one group into another. Thus, if G_1 and G_2 are groups in which the laws of composition are written multiplicatively, a homomorphism of G_1 into G_2 is a mapping ϕ of G_1 into G_2 such that $\phi(ab) = \phi(a)\phi(b)$ for every pair of elements a, b of G_1. The other terms are used in an obvious way.

Example 1. If R_1 and R_2 are any rings, then the mapping ζ of R_1 into R_2 which assigns to every element of R_1 the zero element of R_2 is a homomorphism of R_1 into R_2; we call it the **zero homomorphism**. If R_1 does not consist of the zero element alone ζ is not a monomorphism; if R_2 does not consist of the zero element alone ζ is not an epimorphism.

Example 2. If R is any ring and ε is the mapping of R

into itself defined by setting $\varepsilon(x) = x$ for all elements x of R, then ε is an automorphism of R; we call it the **identity automorphism**.

Example 3. If $R_1 = \mathbf{Z}$ and $R_2 = \mathbf{Z}_m$ for some positive integer m, $m > 1$, then the mapping which assigns to each integer in $\mathbf{Z}$ its residue class modulo m (see Example 4 in § 1) is an epimorphism of $\mathbf{Z}$ onto $\mathbf{Z}_m$. This mapping is clearly not a monomorphism, since any two integers which are congruent modulo m are mapped onto the same element of $\mathbf{Z}_m$.

Example 4. If $R_1 = \mathbf{Z}$ and $R_2 = \mathbf{Q}$, then the mapping which assigns to each integer n the rational number $n/1$ is a monomorphism of $\mathbf{Z}$ into $\mathbf{Q}$, which is obviously not an epimorphism.

Example 5. If $R_1 = R_2 = \mathbf{C}$, then the mapping which assigns to each complex number $a+bi$ the conjugate complex number $a-bi$ is an automorphism of $\mathbf{C}$.

It is easy to see that if ϕ is a homomorphism of a ring R_1, with zero element 0_1, into a ring R_2, with zero element 0_2, then $\phi(0_1) = 0_2$. In fact, if a_2 is any element of $\phi(R_1)$, then there is an element a_1 of R_1 such that $a_2 = \phi(a_1)$ and we have

$$\begin{aligned}\phi(0_1)+a_2 = \phi(0_1)+\phi(a_1) &= \phi(0_1+a_1)\\ &= \phi(a_1) = a_2 = 0_2+a_2\end{aligned}$$

from which the desired result follows. Further, if $a_2 = \phi(a_1)$, we have $-a_2 = \phi(-a_1)$ because

$$\begin{aligned}a_2+\phi(-a_1) = \phi(a_1)+\phi(-a_1)&\\ &= \phi(a_1+(-a_1)) = \phi(0_1) = 0_2.\end{aligned}$$

Similar arguments show that if ϕ is a non-zero homomorphism of a field F_1, with identity e_1, into a field F_2, with identity e_2, then $\phi(e_1) = e_2$ and if $a_2 = \phi(a_1)$ is non-zero, then $a_2^{-1} = \phi(a_1^{-1})$. It follows that if ϕ is a non-zero homomorphism of a field F_1 into a field F_2 then $\phi(F_1)$ is a subfield of F_2.

In the preceding paragraph we used different symbols (0_1 and 0_2) for the zero elements of the rings R_1 and R_2. From now on we shall use the same symbol 0 indiscriminately for the zero element of every ring. This should not give rise to any confusion, since the context will usually make it quite clear to what ring an element denoted by 0 is supposed to belong. For example, if ϕ is a homomorphism of R_1 into R_2 and we write $\phi(0) = 0$, it is plain that the element denoted by 0 on the left is the zero element of R_1 (since $\phi(a)$ is defined only for elements a in R_1) and that the element denoted by 0 on the right is the zero element of R_2 (since it is an image under ϕ).

The **kernel** of a homomorphism ϕ of a ring R_1 into a ring R_2 is the set of elements of R_1 mapped by ϕ onto the zero element of R_2. As we have seen, the zero element of R_1 always belongs to the kernel of ϕ; our first theorem shows that if R_1 is a field and ϕ is not the zero homomorphism ζ then the kernel of ϕ contains no other element of R_1.

THEOREM 3.1. *A homomorphism of a field into any ring is either the zero homomorphism or else a monomorphism.*

Proof. Let ϕ be a homomorphism of a field F into a ring R and suppose that ϕ is not a monomorphism; we shall prove that ϕ must be the zero homomorphism.

Since ϕ is not one-to-one there must be two distinct elements a, b of F such that $\phi(a) = \phi(b)$; set $k = a-b$. Then we have

$$\phi(k) = \phi(a+(-b)) = \phi(a)+\phi(-b) = \phi(a)-\phi(b) = 0.$$

Further, since a and b are distinct, k is non-zero and hence has a multiplicative inverse k^{-1}. Thus if x is any element of F we may write $x = k(k^{-1}x)$ and so

$$\phi(x) = \phi(k(k^{-1}x)) = \phi(k)\phi(k^{-1}x) = 0\phi(k^{-1}x) = 0.$$

Hence ϕ is the zero homomorphism as required.

We hinted in § 2 that the examples of prime fields which we gave there—the rational number field $\mathbf{Q}$ and the fields $\mathbf{Z}_p$—are essentially the only prime fields. We now make this statement quite precise.

THEOREM 3.2. *The prime subfield of a field of characteristic zero is isomorphic to the field of rational numbers* $\mathbf{Q}$. *The prime subfield of a field of non-zero characteristic* p *is isomorphic to the field* $\mathbf{Z}_p$.

Proof. (1) Let F be a field of characteristic zero with identity e and prime subfield P. Since P is a subfield of F it contains e; it then contains also the integral multiples me of e, none of which is zero except $0e$; it follows that it contains the multiplicative inverses $(ne)^{-1}$ for all non-zero integers n. Thus P contains all the elements of F of the form $(me)(ne)^{-1}$ where m and n are integers and n is non-zero. But the subset of F consisting of all elements of this form is a subfield of F: to see this we have only to verify that

$$\begin{aligned}(m_1e)(n_1e)^{-1}+(m_2e)(n_2e)^{-1} &= ((m_1n_2+m_2n_1)e)(n_1n_2e)^{-1},\\ (m_1e)(n_1e)^{-1}\,.\,(m_2e)(n_2e)^{-1} &= (m_1m_2e)(n_1n_2e)^{-1},\\ -(me)(ne)^{-1} &= ((-m)e)(ne)^{-1},\\ ((me)(ne)^{-1})^{-1} &= (ne)(me)^{-1} \text{ whenever } m\neq 0.\end{aligned}$$

It follows that P is made up precisely of the elements of this form. It is now a purely routine matter to verify that the mapping ϕ of $\mathbf{Q}$ into P given by $\phi(m/n) = (me)(ne)^{-1}$ is an isomorphism.

(2) Let now F be a field of non-zero characteristic p with identity e and prime subfield P. Then P contains

the p distinct integral multiples of e: $0e = 0$, $1e = e$, $2e$, $\ldots$, $(p-1)e$.

Let $C_0, C_1, \ldots, C_{p-1}$ be the residue classes of $0, 1, \ldots, p-1$ modulo p; these are, of course, the elements of the field $\mathbf{Z}_p$. Define the mapping ϕ of $\mathbf{Z}_p$ into F by setting $\phi(C_m) = me$ $(m = 0, 1, \ldots, p-1)$. Then ϕ is a homomorphism. For, let C_k and C_l be elements of $\mathbf{Z}_p$ $(0 \leqq k, l < p)$; set $k+l = qp+r$ $(0 \leqq r < p)$; then $C_k + C_l = C_r$ and we have

$$\phi(C_k)+\phi(C_l) = ke+le = (k+l)e = re = \phi(C_r) = \phi(C_k+C_l).$$

A similar argument shows that $\phi(C_k)\phi(C_l) = \phi(C_kC_l)$.

Since ϕ is clearly not the zero homomorphism it is a monomorphism of $\mathbf{Z}_p$ onto the set of integral multiples of e, which is therefore a subfield of F. It follows that P consists precisely of the integral multiples of e; ϕ is then an isomorphism of $\mathbf{Z}_p$ onto P.

Corollary. *Finite fields have non-zero characteristics.*

Proof. Let F be a finite field, P its prime field. If F has characteristic zero, then P is isomorphic to $\mathbf{Q}$ and hence is an infinite field. But there is no room for an infinite subfield in a finite field.

We conclude this section by introducing an interesting and important mapping of fields of non-zero characteristic.

Theorem 3.3. *Let F be a field of non-zero characteristic p. Then the mapping π of F into itself defined by setting $\pi(x) = x^p$ for all elements x of F is a monomorphism.*

Proof. Let a and b be any two elements of F. According to the binomial theorem (which is valid in any field since the multiplication operation is commutative),

$$\pi(a+b) = (a+b)^p = a^p + \sum_{0<k<p} \binom{p}{k} a^{p-k}b^k + b^p,$$

where $\binom{p}{k}$ is the (ordinary integer) binomial coefficient.

Suppose $0<k<p$. Then none of the integers $1, 2, \ldots, k$ is divisible by p and hence their product $k!$ is not divisible by p; similarly $(p-k)!$ is not divisible by p. It follows that $\binom{p}{k} = p!/k!(p-k)!$ is divisible by p and hence that $\binom{p}{k}a^{p-k}b^k = 0$. Thus we have

$$\pi(a+b) = a^p + b^p = \pi(a) + \pi(b).$$

Of course, since multiplication in F is commutative,

$$\pi(ab) = (ab)^p = a^p b^p = \pi(a)\pi(b).$$

Thus π is a homomorphism of F into F. Since $\pi(e) = e^p = e$ is non-zero, π is not the zero homomorphism and hence, by Theorem 3.1, π is a monomorphism as required.

We remark that since π is a homomorphism we have $\pi(a-b) = \pi(a) - \pi(b)$, i.e. $(a-b)^p = a^p - b^p$ for all pairs of elements a, b in F. When $p = 2$ this appears to contradict the natural computation $(a-b)^2 = a^2 - 2ab + b^2 = a^2 + b^2$, until we recall that in a field of characteristic 2 we have $2b^2 = 0$ and hence $b^2 = -b^2$!

A simple inductive argument leads to the following useful consequence of Theorem 3.3.

COROLLARY. *Let F be a field of non-zero characteristic p. Then for every positive integer k the mapping π_k of F into itself defined by setting $\pi_k(x) = x^{p^k}$ for all elements x of F is a monomorphism.*

§ **4. Vector spaces.** We collect in this section a few very elementary results about vector spaces which we shall require in the sequel. Let F be a field; a **vector space** over F is a set V equipped with a law of composition which we shall call addition and an operation which assigns to every element α of F and every element x of V an element of V which we denote by αx, such that the following conditions are satisfied:

A. The elements of V form an abelian group under the addition operation;

M. For all elements α, β of F and all elements x, y of V we have

$$(\alpha+\beta)x = \alpha x+\beta x;$$
$$\alpha(x+y) = \alpha x+\alpha y;$$
$$\alpha(\beta x) = (\alpha\beta)x;$$
$$ex = x \text{ (where } e \text{ is the identity of } F).$$

We remark that two zero elements will occur in our discussions—the zero element of F and the zero element of V. We may denote them by 0_F and 0_V respectively when we want to emphasise which we are dealing with, but in fact there is little danger of confusion if we denote them both simply by 0, and we shall usually do this.

First we note some elementary consequences of the definition of vector spaces. Let x be any element of V; then $0_F x = 0_V$. To prove this, we remark that $0_F+0_F = 0_F$ and so $0_F x+0_F x = (0_F+0_F)x = 0_F x$; adding $-(0_F x)$ to both sides we obtain the desired result. An equally simple argument shows that for every element α of F, $\alpha 0_V = 0_V$. Finally we show that for every element α in F and every element x in V we have $(-\alpha)x = -(\alpha x) = \alpha(-x)$. The first of these results follows when we notice that

$$\alpha x+(-\alpha)x = (\alpha+(-\alpha))x = 0_F x = 0_V$$

and the second from the similar result that

$$\alpha x+\alpha(-x) = \alpha(x+(-x)) = \alpha 0_V = 0_V.$$

If $x_1, \ldots, x_k$ are elements of V, a **linear combination of** $x_1, \ldots, x_k$ **with coefficients in** F is an element of the form $\alpha_1 x_1+\ldots+\alpha_k x_k$, where $\alpha_1, \ldots, \alpha_k$ are elements of F.

Let V be a vector space over a field F. A finite subset $\{x_1, \ldots, x_k\}$ of V is said to be **linearly dependent** over F if there exist elements $\alpha_1, \ldots, \alpha_k$ of F, not all zero, such that

$$\alpha_1 x_1 + \ldots + \alpha_k x_k = 0.$$

Otherwise the subset is said to be **linearly independent** over F. Thus the subset $\{x_1, \ldots, x_k\}$ of V is linearly independent over F if the only linear combination of $x_1, \ldots, x_k$ with coefficients in F which can be zero is that in which all the coefficients are zero; in other words, if the only relation of the form

$$\alpha_1 x_1 + \ldots + \alpha_k x_k = 0$$

(with $\alpha_1, \ldots, \alpha_k$ in F) which can hold is that in which $\alpha_1 = \ldots = \alpha_k = 0$. It is an immediate consequence of the definition that a subset consisting of a single non-zero element x of V is linearly independent. For, suppose we had a relation $\alpha x = 0$, with α non-zero; then α has a multiplicative inverse α^{-1} and we have

$$x = ex = (\alpha^{-1}\alpha)x = \alpha^{-1}(\alpha x) = \alpha^{-1}0 = 0,$$

which is a contradiction. On the other hand, the subset consisting of the zero element of V alone is linearly dependent. An infinite subset of V is said to be linearly independent if all its finite subsets are independent.

The **dimension** of a vector space V over F is the maximum number of linearly independent elements of V (over F). That is to say, if there exists an integer n such that there is a linearly independent subset consisting of n elements of V, while every subset consisting of more than n elements of V is linearly dependent, then the dimension of V over F is n; if there is no such integer n, i.e., if there exist linearly independent subsets with arbitrarily many elements, then the dimension of V over F is infinite. We denote the dimension of V over F by $\dim_F V$.

Our next theorem is an immediate consequence of the definition.

THEOREM 4.1. *If a vector space V over F includes a linearly independent subset consisting of m elements, then* $\dim_F V \geqq m$.

A subset S of a vector space V over F is said to **generate** V over F or to be a **generating system** for V over F if every element of V can be expressed as a linear combination of a finite number of elements of S with coefficients in F; this expression is not required to be unique. A linearly independent generating system for V is called a **basis** for V over F. If $\{x_1, \ldots, x_n\}$ is a basis for V, then the expression of an element x of V as a linear combination of $x_1, \ldots, x_n$ is unique. For, if we have

$$\alpha_1 x_1 + \ldots + \alpha_n x_n = x = \beta_1 x_1 + \ldots + \beta_n x_n,$$

it follows that

$$(\alpha_1 - \beta_1)x_1 + \ldots + (\alpha_n - \beta_n)x_n = 0,$$

and hence, since $\{x_1, \ldots, x_n\}$ is linearly independent, all the coefficients $\alpha_i - \beta_i$ are zero ($i = 1, \ldots, n$), i.e.,

$$\alpha_i = \beta_i \quad (i = 1, \ldots, n).$$

Before stating our next theorem we recall a well-known result from the theory of linear equations on which we make it depend: namely, a system of homogeneous linear equations in which there are more unknowns than equations has a non-trivial solution. We put this more formally, as follows. Let α_{ij} ($i = 1, \ldots, m$; $j = 1, \ldots, n$) be elements of a field F; then, if $n > m$, there exist elements $\beta_1, \ldots, \beta_n$ in F, not all zero, such that

$$\alpha_{i1}\beta_1 + \ldots + \alpha_{in}\beta_n = 0 \quad (i = 1, \ldots, m).$$

THEOREM 4.2. *If a vector space V over F has a generating system consisting of m elements, then* $\dim_F V \leqq m$.

Proof. Let $\{x_1, \ldots, x_m\}$ be a generating system for V.

We shall show that every subset of V containing more than m elements is linearly dependent. So let $\{y_1, ..., y_n\}$ $(n > m)$ be any such subset.

Since $\{x_1, ..., x_m\}$ generates V, there exist elements α_{ij} of F $(i = 1, ..., m;\ j = 1, ..., n)$ such that

$$y_j = \alpha_{1j}x_1 + ... + \alpha_{mj}x_m \quad (j = 1, ..., n).$$

Now if $\beta_1, ..., \beta_n$ are any elements of F we have

$$\beta_1 y_1 + ... + \beta_n y_n = (\alpha_{11}\beta_1 + ... + \alpha_{1n}\beta_n)x_1 + ... + (\alpha_{m1}\beta_1 + ... + \alpha_{mn}\beta_n)x_m.$$

But since $n > m$ it follows from the result about homogeneous linear equations which we quoted above that there are elements $\beta_1, ..., \beta_n$ of F, not all zero, such that

$$\alpha_{i1}\beta_1 + ... + \alpha_{in}\beta_n = 0 \quad (i = 1, ..., m)$$

and hence such that

$$\beta_1 y_1 + ... \beta_n y_n = 0.$$

Thus the subset $\{y_1, ..., y_n\}$ is linearly dependent as we claimed.

It follows at once from Theorems 4.1 and 4.2 that if we want to prove that a vector space V over a field F has dimension m over F, a useful way to proceed will be to establish that V has a basis consisting of m elements. For, by Theorem 4.1, the linear independence of the basis will show that $\dim_F V \geqq m$, while, by Theorem 4.2, the fact that a basis generates V will yield the reversed inequality $\dim_F V \leqq m$. Combining the two inequalities we shall, of course, obtain the result that $\dim_F V = m$.

Example. Let F be any field; let V be the set of ordered m-tuples $(\alpha_1, ..., \alpha_m)$ of elements of F. We define an operation of addition in V by setting

$$(\alpha_1, ..., \alpha_m) + (\beta_1, ..., \beta_m) = (\alpha_1 + \beta_1, ..., \alpha_m + \beta_m)$$

and an operation which assigns to each element α of F and every element $(\alpha_1, \ldots, \alpha_m)$ of V the element

$$\alpha(\alpha_1, \ldots, \alpha_m) = (\alpha\alpha_1, \ldots, \alpha\alpha_m)$$

of V. It is easily verified that, with these two operations, V is a vector space over F.

Let e_i $(i = 1, \ldots, m)$ be the element of V which has the identity element e of F in the ith place and the zero element of F in the remaining places. We claim that $\{e_1, \ldots, e_m\}$ is a basis for V over F. First of all, it is clear that this subset generates V over F; for every element $(\alpha_1, \ldots, \alpha_m)$ of V can be expressed in the form

$$(\alpha_1, \ldots, \alpha_m) = \alpha_1 e_1 + \ldots + \alpha_m e_m.$$

But the set $\{e_1, \ldots, e_m\}$ is also linearly independent; for if $\alpha_1 e_1 + \ldots + \alpha_m e_m = (\alpha_1, \ldots, \alpha_m)$ is the zero element $(0, \ldots, 0)$ of V, then all the coefficients α_i must be zero $(i = 1, \ldots, m)$. Hence $\{e_1, \ldots, e_m\}$ is a basis for V over F and so $\dim_F V = m$.

THEOREM 4.3. *Let V be a vector space of finite dimension m over a field F. Then* (1) *every linearly independent subset of V consisting of m elements is a basis for V over F, and* (2) *every generating system for V over F consisting of m elements is a basis for V over F.*

Proof. (1) Let $\{x_1, \ldots, x_m\}$ be a linearly independent subset of V consisting of m elements. We shall show that this subset is also a generating system for V over F, and hence is a basis.

Clearly each of the elements x_i $(i = 1, \ldots, m)$ is a linear combination of $x_1, \ldots, x_m$. Let now x be any other element of V; then the subset $\{x, x_1, \ldots, x_m\}$ consists of $m+1$ elements and hence cannot be linearly independent. So there exist elements $\alpha, \alpha_1, \ldots, \alpha_m$ of F, not all zero, such that

$$\alpha x + \alpha_1 x_1 + \ldots + \alpha_m x_m = 0. \tag{4.1}$$

It is clear that α is non-zero; for, if α were zero, it would follow at once from (4.1) that $\{x_1, ..., x_m\}$ is linearly dependent, contrary to hypothesis. Hence α has a multiplicative inverse, and we deduce that

$$x = (-\alpha^{-1}\alpha_1)x_1 + ... + (-\alpha^{-1}\alpha_m)x_m.$$

Thus $\{x_1, ..., x_m\}$ generates V, as asserted.

(2) Now let $\{y_1, ..., y_m\}$ be a generating system for V over F consisting of m elements. We shall show that this subset is linearly independent over F and hence is a basis.

Suppose that the subset is linearly dependent over F; then there exist elements $\beta_1, ..., \beta_m$ of F, not all zero, such that

$$\beta_1 y_1 + ... + \beta_m y_m = 0.$$

Say β_m is non-zero. Then we may write

$$y_m = (-\beta_m^{-1}\beta_1)y_1 + ... + (-\beta_m^{-1}\beta_{m-1})y_{m-1}. \quad (4.2)$$

Now, since $\{y_1, ..., y_m\}$ is a generating system for V over F, every element of V can be expressed as a linear combination of $y_1, ..., y_m$ with coefficients in F. But (4.2) shows that every such linear combination can be transformed into a linear combination of $y_1, ..., y_{m-1}$ with coefficients in F; that is to say, $\{y_1, ..., y_{m-1}\}$ generates V over F. It follows from Theorem 4.2 that $m = \dim_F V \leq m-1$, which is a contradiction.

Thus $\{y_1, ..., y_m\}$ is linearly independent over F, as asserted.

Since, by definition, every vector space of finite dimension m certainly includes a linearly independent subset consisting of m elements, it follows at once from the first part of Theorem 4.3 that every vector space of finite dimension m has a basis consisting of m elements.

Finally we show that every linearly independent subset of a finite-dimensional vector space can be extended to a basis for the vector space.

THEOREM 4.4. *Let V be a vector space of finite dimension m over a field F. If $\{x_1, ..., x_r\}$ is a linearly independent subset of V, $r \leq m$, then there exist $m-r$ elements $x_{r+1}, ..., x_m$ of V such that $\{x_1, ..., x_m\}$ is a basis for V over F.*

Proof. If $r = m$ there is nothing to prove.

So suppose that $r<m$. The subset W of V consisting of all elements of the form $\alpha_1 x_1 + ... + \alpha_r x_r$ (where $\alpha_1, ..., \alpha_r$ are elements of F) is easily seen to be a vector space. Since it has a linearly independent generating system consisting of r elements—namely $\{x_1, ..., x_r\}$—its dimension is r. Hence W is not the whole space V. Let x_{r+1} be any element of V which does not belong to W; we claim that $\{x_1, ..., x_{r+1}\}$ is linearly independent. So suppose we have

$$\beta_1 x_1 + ... + \beta_r x_r + \beta_{r+1} x_{r+1} = 0,$$

where $\beta_1, ..., \beta_{r+1}$ are elements of F. If β_{r+1} is non-zero, then it has a multiplicative inverse and we may write

$$x_{r+1} = (-\beta_{r+1}^{-1}\beta_1)x_1 + ... + (-\beta_{r+1}^{-1}\beta_r)x_r,$$

which implies that x_{r+1} lies in W, contrary to hypothesis. Thus $\beta_{r+1} = 0$ and the above relation becomes

$$\beta_1 x_1 + ... + \beta_r x_r = 0.$$

Hence $\beta_1 = ... = \beta_r = 0$, since $\{x_1, ..., x_r\}$ is linearly independent. Thus $\{x_1, ..., x_{r+1}\}$ is linearly independent. Repeating this procedure $m-r$ times in all, we eventually produce a linearly independent subset

$$\{x_1, ..., x_r, x_{r+1}, ..., x_m\}$$

consisting of m elements. Theorem 4.3 shows that this subset is a basis for V over F as required.

§ 5. **Polynomials.** In elementary books on algebra, polynomials are usually defined to be " expressions of the form

$$f(x) \equiv a_0 + a_1x + a_2x^2 + \ldots + a_nx^n$$

where a_0, ..., a_n are numbers ". This definition is open to at least two objections. First, it gives no indication as to the logical status of the symbol x—and to say that x is a " variable " or an " indeterminate " simply begs further questions. Secondly, when one comes to define addition and multiplication of polynomials, it is difficult to avoid the feeling that these operations are at least partially defined already—for the polynomials are written with addition signs between their terms, and these terms contain powers of x. To avoid these objections we proceed in what may appear to be a more abstract way; in reality we are simply exploiting the well-known fact that in elementary algebra the powers of x act essentially only as " place-holders " while the coefficients are the really important constituents. These remarks may already have suggested to the sophisticated reader that we might define polynomials to be simply finite sequences of coefficients such as $(a_0, a_1, \ldots, a_n)$. For technical reasons—in fact, to enable us to deal conveniently with polynomials of different degrees, which would correspond to sequences of different lengths—we prefer to deal with " essentially finite " infinite sequences. We now proceed with the formal development.

Let R be a commutative ring with identity element e. We denote by $P(R)$ the set of infinite sequences $(a_0, a_1, \ldots, a_n, \ldots)$ of elements of R, each of which has the property that only finitely many of the members a_i of the sequence are non-zero; thus for each sequence $a = (a_0, a_1, a_2, \ldots)$ in $P(R)$ there is an integer N_a such that $a_i = 0$ for all integers $i > N_a$. It is important to be clear that two sequences are equal if and only if corresponding members are equal, i.e. if $a = (a_0, a_1, a_2, \ldots)$ and $b = (b_0, b_1, b_2, \ldots)$ then $a = b$ if and only if $a_i = b_i$ $(i = 0, 1, 2, \ldots)$.

We introduce an operation of addition in $P(R)$ by setting

$$(a_0, a_1, a_2, \ldots)+(b_0, b_1, b_2, \ldots)$$
$$= (a_0+b_0, a_1+b_1, a_2+b_2, \ldots).$$

It is easily verified that under this law of composition $P(R)$ forms an abelian group. The zero element is clearly the sequence $z = (0, 0, 0, \ldots)$ each of whose members is the zero of R; the additive inverse of $(a_0, a_1, a_2, \ldots)$ is $(-a_0, -a_1, -a_2, \ldots)$.

Next we introduce an operation of multiplication in $P(R)$ by setting

$$(a_0, a_1, a_2, \ldots)(b_0, b_1, b_2, \ldots) = (c_0, c_1, c_2, \ldots),$$

where

$$c_n = \sum_{i=0}^{n} a_i b_{n-i} \quad (n = 0, 1, 2, \ldots).$$

Then it is a routine matter to verify that this multiplication is associative and commutative, and also distributive with respect to the addition. That is to say, $P(R)$ is a commutative ring under the laws of composition we have defined. Further, $P(R)$ has an identity element, namely the sequence $(e, 0, 0, \ldots)$ where e is the identity element of R.

Consider now the mapping κ of R into $P(R)$ defined by setting $\kappa(a_0) = (a_0, 0, 0, \ldots)$ for all elements a_0 of R. It is easy to see that κ is a monomorphism; we call it the **canonical monomorphism** of R into $P(R)$. Then R and its image $\kappa(R)$ under κ are isomorphic: they differ, of course, in the nature of their elements, but have exactly the same structure. We frequently find it convenient to blur the distinction between R and $\kappa(R)$ and to use the same symbol a_0 both for an element of R and for its image under κ in $P(R)$; when we do this, we say that we are **identifying**

R with its image under κ and regarding R as a subring of the ring $P(R)$. It will be found in practice that very little confusion is likely to arise from this identification procedure; but any confusion which does arise can be resolved by a return to the strictly logical notation.

We now introduce a name for the special sequence $(0, e, 0, 0, \ldots)$ in $P(R)$: we call it X. By induction we can prove at once that, for every positive integer n, X^n is the sequence $(c_0, c_1, c_2, \ldots)$ for which $c_n = e$ and $c_i = 0$ whenever $i \neq n$. Then if $f = (a_0, a_1, \ldots, a_N, 0, 0, \ldots)$ is any sequence in $P(R)$, with $a_n = 0$ for all integers $n > N$, we have

$$
\begin{aligned}
f &= (a_0, 0, 0, \ldots)+(0, a_1, 0, \ldots)+\ldots+(0, \ldots, 0, a_N, 0, \ldots) \\
&= \kappa(a_0)+\kappa(a_1)X+\ldots+\kappa(a_N)X^N. \qquad (5.1)
\end{aligned}
$$

Carrying out the identification of R and $\kappa(R)$ described in the last paragraph, we see that we have expressed f in the form

$$f = a_0+a_1X+\ldots+a_NX^N.$$

This provides the justification for calling the elements of $P(R)$ **polynomials** and for describing $P(R)$ as the **ring of polynomials with coefficients in** R.

Let f be a non-zero polynomial with coefficients in R: say $f = (a_0, a_1, a_2, \ldots)$. We define the **degree** of f to be the greatest integer n such that a_n is non-zero; we denote the degree of f by ∂f. The polynomials of degree zero are precisely the non-zero elements of the subring $\kappa(R)$; we call them **constant polynomials** or simply **constants**. Polynomials of degree 1 are also called **linear polynomials**. It is convenient to define the degree of the zero polynomial $z = \kappa(0)$ to be $-\infty$, with the usual understanding that for every integer $n \geqq 0$ we have $n > -\infty$ and $-\infty+n = -\infty$. We deduce immediately from the definitions of addition and multiplication that if f and g are polynomials with coefficients in R, then

$$\partial(f+g) \leqq \max(\partial f, \partial g)$$

and

$$\partial(fg) \leqq \partial f + \partial g;$$

we notice, too, that if $\partial f \neq \partial g$ then we actually have $\partial(f+g) = \max(\partial f, \partial g)$ and that if R is a field then

$$\partial(fg) = \partial f + \partial g.$$

If $f = (a_0, a_1, \ldots, a_N, \ldots)$ is a non-zero polynomial with degree N, we call a_N the **leading coefficient** of f; this name perhaps appears more reasonable when we express f in the form $f = a_N X^N + \ldots + a_1 X + a_0$. If the leading coefficient of f is the identity e of R, we say that f is a **monic polynomial**, and we drop the leading coefficient, writing simply $f = X^N + \ldots + a_1 X + a_0$.

We now concentrate our attention on polynomials with coefficients in a field. So let F be a field; a polynomial f with coefficients in F is said to be **divisible** by another such polynomial d, and d is said to be a **factor** of f, if there exists a polynomial q such that $f = qd$. In this situation we say also that f is a **multiple** of d. The polynomial f is said to be **irreducible** if it has no factor d such that $0 < \partial d < \partial f$; thus the only factors of an irreducible polynomial f are the constant polynomials and the products of f by the constant polynomials.

We now state without proof two theorems to which we shall have constant recourse in the next chapter. Proofs of these results may be found in many books on elementary algebra, where they are often described as the Division Algorithm and the Euclidean Algorithm respectively.

THEOREM 5.1. *Let f be any polynomial and let d be a non-zero polynomial with coefficients in F. Then there exist unique polynomials q and r with coefficients in F such that $f = qd + r$ and $\partial r < \partial d$.*

THEOREM 5.2. *Let f and g be any two non-zero polynomials with coefficients in F. Then there exists a unique monic polynomial h with coefficients in F such that* (1) *h is a factor of both f and g;* (2) *if k is any polynomial which is a factor of both f and g, then k is a factor of h. Further, there exist polynomials a and b with coefficients in F such that $h = af+bg$.*

The polynomials q and r in Theorem 5.1 are called respectively the **quotient** and **remainder** when f is divided by d. The unique polynomial h described in Theorem 5.2 is called the **highest common factor** or **greatest common divisor** of f and g. If the highest common factor of f and g is the constant polynomial e (to be quite precise we should call it $\kappa(e)$), we say that f and g are **relatively prime.**

We now introduce a very important mapping of the polynomial ring $P(F)$ into itself; this is the mapping D defined by setting

$$Df = D(a_0+a_1X+a_2X^2+\ldots+a_nX^n)$$
$$= a_1+2a_2X+3a_3X^2+\ldots+na_nX^{n-1}$$

for every polynomial $f = a_0+a_1X+a_2X^2+\ldots+a_nX^n$ in $P(F)$. We naturally call the polynomial Df the **derivative** of the polynomial f. It is no surprise to learn that, if f and g are any two polynomials in $P(F)$, then

$$D(f+g) = Df+Dg \text{ and } D(fg) = (Df)g+f(Dg).$$

Nor is it difficult to prove these results, which are purely formal consequences of the definition of D.

It is, of course, quite impracticable to define derivatives of polynomials with coefficients in a general field F by the familiar type of limiting process used in the calculus; for one reason, we have not defined polynomials as functions; for another, in a general field we do not have any notion of " limit ". But the definition we have given makes sense in any field F, since the coefficients $2a_2$, $3a_3$, ... are integral multiples of elements of F and hence are well-determined elements of F (see § 2).

The next result is also an immediate consequence of the definition.

THEOREM 5.3. *Let $f = a_0 + a_1X + \ldots + a_nX^n$ be a polynomial in $P(F)$. If F has characteristic zero then $Df = z$ (the zero polynomial) if and only if f is either zero or a constant polynomial, i.e., if and only if $a_1 = a_2 = \ldots = a_n = 0$. If f has non-zero characteristic p then $Df = z$ if and only if $a_k = 0$ for all integers k not divisible by p.*

Let R be any commutative ring including the field F, and let α be any element of R. An element β of R which can be expressed (not necessarily uniquely) in the form $\beta = a_0 + a_1\alpha + a_2\alpha^2 + \ldots + a_n\alpha^n$, where a_0, a_1, ..., a_n are elements of F and n is a non-negative integer, is called a **polynomial in α with coefficients in** F. The set of all such elements is a subring of R, which we denote by $F[\alpha]$. If κ denotes, as before, the canonical monomorphism of F into $P(F)$, then equation (5.1) shows that $P(F) = \kappa(F)[X]$ and so, identifying F and $\kappa(F)$, we have $P(F) = F[X]$ which is a standard notation for the polynomial ring with coefficients in F.

Returning to the general case, we define a mapping σ_α of the polynomial ring $P(F)$ into R by setting

$$\sigma_\alpha(f) = \sigma_\alpha(a_0 + a_1X + \ldots + a_nX^n) = a_0 + a_1\alpha + \ldots + a_n\alpha^n$$

for every polynomial $f = a_0 + a_1X + \ldots + a_nX^n$ in $P(F)$. We call σ_α the operation of **substituting** α for X; clearly σ_α maps $P(F)$ onto the subring $F[\alpha]$ of R. But we can say more than this, as follows.

THEOREM 5.4. *If R is a commutative ring including the field F and α is any element of R then the mapping σ_α is an epimorphism of $P(F)$ onto $F[\alpha]$.*

Proof. Let

$$f = a_0+a_1X+a_2X^2+\dots \text{ and } g = b_0+b_1X+b_2X^2+\dots$$

be any two polynomials in $P(F)$. Then

$$f+g = (a_0+b_0)+(a_1+b_1)X+(a_2+b_2)X^2+\dots$$

and

$$fg = (a_0b_0)+(a_0b_1+a_1b_0)X+(a_0b_2+a_1b_1+a_2b_0)X^2+\dots.$$

We see at once that

$$\begin{aligned}\sigma_\alpha(f)+\sigma_\alpha(g) &= (a_0+a_1\alpha+a_2\alpha^2+\dots)+(b_0+b_1\alpha+b_2\alpha^2+\dots)\\ &= (a_0+b_0)+(a_1+b_1)\alpha+(a_2+b_2)\alpha^2+\dots\\ &= \sigma_\alpha(f+g).\end{aligned}$$

We have also

$$\begin{aligned}&\sigma_\alpha(f)\sigma_\alpha(g)\\ &\quad= (a_0+a_1\alpha+a_2\alpha^2+\dots)(b_0+b_1\alpha+b_2\alpha^2+\dots)\\ &\quad= a_0b_0+(a_0b_1\alpha+a_1\alpha b_0)\\ &\qquad\qquad+(a_0b_2\alpha^2+a_1\alpha b_1\alpha+a_2\alpha^2b_0)+\dots \quad (5.2)\\ &\quad= a_0b_0+(a_0b_1+a_1b_0)\alpha+(a_0b_2+a_1b_1+a_2b_0)\alpha^2+\dots \quad (5.3)\\ &\quad= \sigma_\alpha(fg).\end{aligned}$$

Thus σ_α is a homomorphism. We remark that we have made essential use of the commutative property of R in passing from (5.2) to (5.3).

If f is a polynomial in $P(F)$ and $\sigma_\alpha(f) = 0$, then we say that α is a **root** of f in R. As in elementary algebra, there is an intimate connexion between the roots of f in the field F itself and the linear factors of f in $P(F)$. If α is any element of F we denote the linear polynomial $X-\alpha$ by l_α.

THEOREM 5.5. *If α is an element of F and f is a polynomial in $P(F)$, then α is a root of f in F if and only if l_α is a factor of f in $P(F)$.*

Proof. According to Theorem 5.1 there exist polynomials q and r in $P(F)$ such that $f = ql_\alpha+r$ and $\partial r<\partial l_\alpha = 1$.

Thus r has the form $\kappa(a)$ where a is an element of F, possibly zero. Then

$$\sigma_\alpha(f) = \sigma_\alpha(ql_\alpha + r) = \sigma_\alpha(q)\sigma_\alpha(l_\alpha) + \sigma_\alpha(r) = \sigma_\alpha(r) = a,$$

since $\sigma_\alpha(l_\alpha) = \alpha - \alpha = 0$. Hence α is a root of f if and only if $a = 0$, i.e., if and only if r is the zero polynomial and so l_α is a factor of f.

We now contend that if two fields are isomorphic then the rings of polynomials with coefficients in those fields are also isomorphic. This result is an easy consequence of the following theorem.

THEOREM 5.6. *Let τ be a monomorphism of a field F_1 into a field F_2; let κ_1, κ_2 be the canonical monomorphisms of F_1 into $P(F_1)$, F_2 into $P(F_2)$ respectively. Then there exists a monomorphism τ_P of $P(F_1)$ into $P(F_2)$ such that for every element a of F_1 we have $\tau_P(\kappa_1(a)) = \kappa_2(\tau(a))$.*

Proof. The mapping τ_P of $P(F_1)$ into $P(F_2)$ defined by setting

$$\tau_P(f) = \tau_P(a_0 + a_1X + \dots + a_nX^n)$$
$$= \tau(a_0) + \tau(a_1)X + \dots + \tau(a_n)X^n$$

for every polynomial $f = a_0 + a_1X + \dots + a_nX^n$ is easily seen to satisfy our requirements.

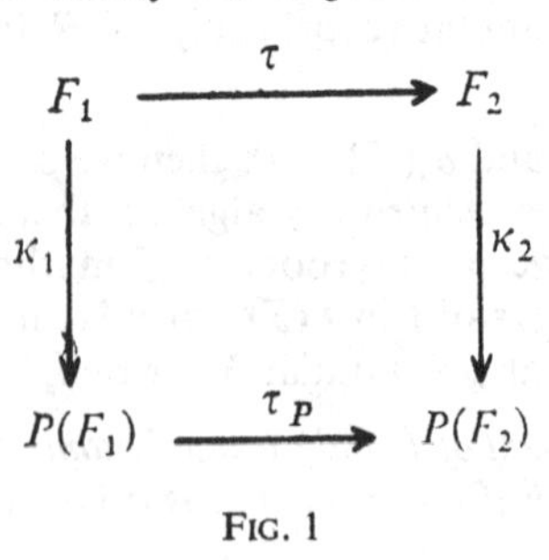

FIG. 1

The condition that $\tau_P(\kappa_1(a)) = \kappa_2(\tau(a))$ for every element a of F_1 is sometimes described by saying that the diagram in fig. 1 is **commutative**; for the condition asserts that if we start with any element a of F_1 and "transport it" to $P(F_2)$ by either of the routes indicated in the diagram —by applying first κ_1 and then τ_P or by applying first τ and then κ_2—we obtain the same result.

If we identify F_1 and F_2 with their images under the canonical monomorphisms and regard them as subfields of $P(F_1)$ and $P(F_2)$ respectively, then the condition on τ_P is simply that $\tau_P(a) = \tau(a)$ for every element a of F_1, i.e. that τ_P shall act like τ on the elements of F_1. For this reason we call τ_P the **canonical extension** of τ to $P(F_1)$.

§ **6. Higher polynomial rings; rational functions.** Let R be any commutative ring with identity e. We define inductively a family of " higher polynomial rings " with coefficients in R, as follows: $P_1(R)$ is simply the polynomial ring $P(R)$ as we defined it in the last section; then for $n>1$, we set $P_n(R) = P(P_{n-1}(R))$. We call $P_n(R)$ the ***n*th order polynomial ring with coefficients in R.**

In order to achieve some insight into the structure of these rings we shall examine $P_2(R) = P(P(R))$. Let κ_1 and κ_2 be the canonical monomorphisms of R into $P(R)$ and of $P(R)$ into $P_2(R)$ respectively; the mapping κ of R into $P_2(R)$ defined by setting $\kappa(a) = \kappa_2(\kappa_1(a))$ for all elements a of R is clearly also a monomorphism. If we denote by X the element $(0, e, 0, ...)$ of $P(R)$, then, as we saw in § 5, every element b of $P(R)$ can be expressed in the form

$$b = \kappa_1(a_0)+\kappa_1(a_1)X+...+\kappa_1(a_m)X^m$$

where $a_0, a_1, ..., a_m$ are elements of R. Similarly, if we denote by X_2 the element $(\kappa_1(0), \kappa_1(e), \kappa_1(0), ...)$ of $P_2(R)$, we see that every element p of $P_2(R)$ can be expressed in the form

$$p = \kappa_2(b_0)+\kappa_2(b_1)X_2+...+\kappa_2(b_n)X_2^n$$

where $b_0, b_1, ..., b_n$ are elements of $P(R)$. If we write

$$b_j = \kappa_1(a_{0j})+\kappa_1(a_{1j})X+...+\kappa_1(a_{m_jj})X^{m_j} \quad (j = 0, 1, ..., n)$$

and set $X_1 = \kappa_2(X)$, then we see that p takes the form

$$\begin{aligned} p = {} & \kappa(a_{00})+\kappa(a_{10})X_1+\ldots+\kappa(a_{m_00})X_1^{m_0} \\ & +(\kappa(a_{01})+\kappa(a_{11})X_1+\ldots+\kappa(a_{m_11})X_1^{m_1})X_2 \\ & +\ldots+ \\ & +(\kappa(a_{0n})+\kappa(a_{1n})X_1+\ldots+\kappa(a_{m_nn})X_1^{m_n})X_2^n \\ = {} & \sum_{j=0}^{n}\sum_{i=0}^{m_j}\kappa(a_{ij})X_1^iX_2^j. \end{aligned}$$

If we identify the original ring R with its image under the monomorphism κ, it appears that every element p of $P_2(R)$ can be expressed in the form

$$p = \sum_{i=0}^{\infty}\sum_{j=0}^{\infty} a_{ij}X_1^iX_2^j$$

where the coefficients a_{ij} belong to the ring R and only finitely many of them are non-zero.

A similar discussion shows that by suitable choice of elements $X_1, \ldots, X_n$ and the usual identification procedure, every element of $P_n(R)$ can be expressed in the form

$$\sum_{i_1=0}^{\infty}\cdots\sum_{i_n=0}^{\infty} a_{i_1\ldots i_n}X_1^{i_1}\ldots X_n^{i_n}$$

where the coefficients $a_{i_1\ldots i_n}$ are in R and only finitely many of them are non-zero.

Let now S be any commutative ring including R and let $\boldsymbol{\alpha} = (\alpha_1, \ldots, \alpha_n)$ be an ordered n-tuple of elements of S. We may then define a mapping $\sigma_{\boldsymbol{\alpha}}$ of $P_n(R)$ into S by setting

$$\sigma_{\boldsymbol{\alpha}}(\sum a_{i_1\ldots i_n}X_1^{i_1}\ldots X_n^{i_n}) = \sum a_{i_1\ldots i_n}\alpha_1^{i_1}\ldots\alpha_n^{i_n}$$

for every element $\sum a_{i_1\ldots i_n}X_1^{i_1}\ldots X_n^{i_n}$ in $P_n(R)$. As in Theorem 5.4 we may prove that $\sigma_{\boldsymbol{\alpha}}$ is a homomorphism. The image of $P_n(R)$ under the mapping $\sigma_{\boldsymbol{\alpha}}$ is a subring of S which is often denoted by $R[\boldsymbol{\alpha}]$ or $R[\alpha_1, ..., \alpha_n]$. In particular, $P_n(R) = R[X_1, ..., X_n]$.

Once again let F be any field and let κ be the canonical monomorphism of F into the polynomial ring $P(F)$. Let us consider the set of ordered pairs (f, g) of polynomials in $P(F)$, where g is not the zero polynomial. We say that the ordered pair (f', g') is equivalent to the ordered pair (f, g) if $f'g = fg'$. The set of all ordered pairs equivalent to (f, g) is called the equivalence class of (f, g); it is clearly not empty, since (f, g) is equivalent to itself under our definition, and hence belongs to the equivalence class of (f, g). The set of all equivalence classes of ordered pairs obtained in this way is denoted by $R(F)$. We turn $R(F)$ into a ring by defining operations of addition and multiplication as follows. Let C_1 and C_2 be any two equivalence classes; let (f_1, g_1) and (f_2, g_2) be ordered pairs of polynomials chosen from C_1 and C_2 respectively. Then we define $C_1 + C_2$ and C_1C_2 to be the equivalence classes containing the ordered pairs $(f_1g_2 + f_2g_1, g_1g_2)$ and $(f_1 f_2, g_1g_2)$ respectively. It is easily verified that these equivalence classes depend only on C_1 and C_2 and not on the choice of ordered pairs (f_1, g_1) and (f_2, g_2). Routine checking now shows that with these laws of composition $R(F)$ is a commutative ring with identity; the zero and identity elements are the equivalence classes containing the pairs $(\kappa(0), \kappa(e))$ and $(\kappa(e), \kappa(e))$ respectively. Let now C be any non-zero element of $R(F)$, and let (f, g) be any pair in C; since C is not the zero element it follows that f is not the zero polynomial. Let C' be the equivalence class containing the pair (g, f); then CC' is the equivalence class containing (fg, gf). Since (fg, gf) is equivalent to $(\kappa(e), \kappa(e))$, it follows that CC' is the identity of $R(F)$. Thus we have proved that every non-zero element of $R(F)$ has a multiplicative inverse, i.e., that $R(F)$ is a field.

We usually denote the equivalence class containing (f, g) by f/g, a notation which perhaps makes the definition of addition and multiplication in $R(F)$ appear more natural. We call $R(F)$ the **field of rational functions with coefficients in F.**

The mapping λ of $P(F)$ into $R(F)$ defined by setting $\lambda(f) = f/\kappa(e)$ for all polynomials f in $P(F)$ is easily seen to be a monomorphism; we call λ the **canonical monomorphism** of $P(F)$ into $R(F)$. Usually we agree to identify $P(F)$ with its image under λ, just as we identify F with its image under κ. So we generally think of F and $P(F)$ as subrings of the field $R(F)$.

Examples I

1. Show that the set of complex numbers of the form $a+bi$ where a and b are rational numbers is a subfield of the complex number field **C**.

2. Show that the set consisting of the four elements $0, e, a, b$ with operations of addition and multiplication defined by means of the tables below is a field of characteristic 2. What is the prime field?

$+$	0	e	a	b
0	0	e	a	b
e	e	0	b	a
a	a	b	0	e
b	b	a	e	0

$\times$	0	e	a	b
0	0	0	0	0
e	0	e	a	b
a	0	a	b	e
b	0	b	e	a

3. Let F be a field, $M_n(F)$ the set of $n \times n$ matrices with elements in F. Show that $M_n(F)$ forms a vector space over F under ordinary matrix addition and the operation which assigns to each element a of F and each matrix $A = [a_{ij}]$ in $M_n(F)$ the matrix $aA = [aa_{ij}]$. Show that the set of n^2 matrices which have the identity element of F in one position and the zero element in all other positions is a basis for $M_n(F)$ over F.

4. Let F be a field, P_n the subset of the polynomial ring $P(F)$ consisting of polynomials f such that $\partial f<n$. Show that P_n forms a vector space over F under ordinary polynomial addition and the operation which assigns to each element a of F and each polynomial

$$f = a_0+a_1X+\ldots+a_{n-1}X^{n-1}$$

the polynomial $af = aa_0+aa_1X+\ldots+aa_{n-1}X^{n-1}$. What is the dimension of P_n over F?

5. Let F be a field. For each polynomial f in $P(F)$ we may define a mapping f^* of F into itself by setting $f^*(\alpha) = \sigma_\alpha(f)$ for every element α in F. Show that f^* is not in general a homomorphism.

6. Let $F = \mathbf{Z}_p$ and let f be the polynomial X^p-X in $P(F)$. Show that while f is not the zero polynomial the mapping f^* of F into itself determined by f as in Example 5 is the zero mapping, i.e., $f^*(\alpha) = 0$ for every element α of F.

7. Let ϕ be a homomorphism of a ring R into a ring S. Show that if a and b belong to the kernel K of ϕ and r is any element of R then $a+b$, $a-b$ and ra all belong to K. Deduce that if $R = P(F)$, the ring of polynomials with coefficients in a field F, and f is a polynomial of least degree in K, then K consists precisely of the multiples of f.

8. Let F be a field. Show that the only elements of the polynomial ring $P(F)$ which have multiplicative inverses are the constant polynomials.

9. Let F be a field, κ the canonical monomorphism of F into the polynomial ring $P(F)$. If ϕ is an automorphism of $P(F)$ show that there is an automorphism ϕ_1 of F such that $\phi(\kappa(a)) = \kappa(\phi_1(a))$ for all elements a of F and that there are elements c, d of F $(c \neq 0)$ such that $\phi(X) = cX+d$.

10. If f is any polynomial of degree n with coefficients in a field F of characteristic zero with identity element e, prove that (in the notation of example 5)

$$f = \sum_{r=0}^{n} (r!e)^{-1}(D^r f)^*(0),$$

where the polynomials $D^r f$ are defined inductively by setting $D^0 f = f$ and $D^r f = D(D^{r-1} f)$ for $r \geqq 1$.

CHAPTER II

EXTENSIONS OF FIELDS

§ 7. **Elementary properties.** Let F be a field. An **extension** of F is a pair (E, ι) consisting of a field E and a monomorphism ι of F into E. If (E, ι) is an extension of F the field E has a subfield, namely $\iota(F)$, which is isomorphic to F. We shall usually identify F with its image under ι and regard F as being itself a subfield of E; when we make this identification we shall simply say that the field E is an extension of F and shall omit all mention of the monomorphism ι. In any case where confusion is likely to arise from this identification procedure we shall of course revert to the formal definition and notation.

Example 1. Let F be a subfield of a field E. There may be many monomorphisms of F into E, but there is one which appears to be particularly " natural ". This is the mapping ι of F into E defined by setting $\iota(a) = a$ for all elements a of F; it is clearly a monomorphism, which we call the **inclusion monomorphism** of F in E. Then (E, ι) is an extension of F according to our definition. We notice that in this case $\iota(F) = F$ and hence there is no danger involved in identifying F and $\iota(F)$. In future, if F is a subfield of a field E and we refer to E as an extension of F it will always be understood that the monomorphism of the extension is the inclusion monomorphism.

Example 2. The elements of the field $\mathbf{C}$ of complex numbers are usually defined to be ordered pairs of real numbers, and the laws of composition in **C** are given by

$$(a, b)+(c, d) = (a+c, b+d)$$

$$(a, b)\times(c, d) = (ac-bd, ad+bc).$$

When **C** is defined in this way it is clear that the field **R** of real numbers is not a subfield of **C**. But the mapping ι given by $\iota(a) = (a, 0)$ for all real numbers a is easily verified to be a monomorphism of **R** into **C**; hence (**C**, ι) is an extension of **R**. In this case the procedure of identifying **R** with its image under ι is a very familiar one.

Let F be a subfield of a field E; E is of course equipped with an addition operation—the field addition; the field multiplication in E is an operation which assigns to every element a of F and every element x of E an element ax of E. It is easily verified that E, equipped with these operations, is a vector space over F. The dimension of E over F when considered as a vector space in this way is called the **degree** of E over F and is denoted by $(E: F)$. If (E, ι) is an extension of a field F we define the degree of this extension over F to be the degree of E over its subfield $\iota(F)$. An extension of F is said to be **finite** or **infinite** according as its degree over F is finite or infinite.

Our first result is essentially an elementary exercise illustrating the ideas of §§ 3 and 4.

THEOREM 7.1. *Let F be a subfield of a field E. If τ is a monomorphism of E into a field K, then* $(\tau(E): \tau(F)) = (E: F)$.

Proof. Let $\{x_1, ..., x_r\}$ be a finite subset of E linearly independent over F. We claim that the subset

$$\{\tau(x_1), ..., \tau(x_r)\}$$

is linearly independent over $\tau(F)$. So suppose

$$\tau(a_1)\tau(x_1)+...+\tau(a_r)\tau(x_r) = 0, \tag{7.1}$$

where $a_1, \ldots, a_r$ are elements of F. Since τ is a homomorphism, this relation may be written in the form

$$\tau(a_1x_1 + \ldots + a_rx_r) = 0;$$

from this we deduce (since τ is a monomorphism) that

$$a_1x_1 + \ldots a_rx_r = 0. \tag{7.2}$$

Since $\{x_1, \ldots, x_r\}$ is linearly independent it follows that all the coefficients a_i in (7.2) are zero. Hence all the coefficients $\tau(a_i)$ in (7.1) are zero, and the set $\{\tau(x_1), \ldots, \tau(x_r)\}$ is linearly independent over $\tau(F)$, as asserted.

This completes the proof if $(E: F)$ is infinite. For then we may take r arbitrarily large, so obtaining subsets of $\tau(E)$ which are linearly independent over $\tau(F)$ and contain arbitrarily many elements.

Suppose now that $(E: F)$ is finite. We may take $r = (E: F)$ and assume that $\{x_1, \ldots, x_r\}$ is a basis for E over F. We shall show that $\{\tau(x_1), \ldots, \tau(x_r)\}$ is a basis for $\tau(E)$ over $\tau(F)$; since the preceding argument shows that this set is linearly independent, we have only to prove that it is a generating system for $\tau(E)$ over $\tau(F)$.

So let $\tau(x)$ be any element of $\tau(E)$, where x is an element of E. The set $\{x_1, \ldots, x_r\}$ generates E over F; so there exist elements $a_1, \ldots, a_r$ of F such that

$$x = a_1x_1 + \ldots + a_rx_r.$$

Applying the monomorphism τ we see that

$$\tau(x) = \tau(a_1)\tau(x_1) + \ldots + \tau(a_r)\tau(x_r);$$

this shows that $\{\tau(x_1), \ldots, \tau(x_r)\}$ generates $\tau(E)$ over $\tau(F)$.
This completes the proof.

It is easy to see that if $E = F$ then $(E: F) = 1$. For the set consisting of the identity e of F alone is linearly independent over F; and it generates F over F, since every element a of F can be expressed in the form $a = ae$.

Conversely, let F be a subfield of E such that $(E: F) = 1$. We shall show that in this case $E = F$. Let e be the identity of E; then e lies in the subfield F. The set consisting of e alone is linearly independent over F and hence, by Theorem 4.3, this set is a basis for E over F. Thus, if x is any element of E there is an element a of F such that $x = ae$. So x, being the product of two elements of F, belongs to F. Hence $E = F$.

If F is a subfield of E and E is a subfield of another field, K, then F is also a subfield of K and we may consider the three degrees $(E: F)$, $(K: E)$ and $(K: F)$. The next theorem establishes an important connexion between them.

THEOREM 7.2. *If F is a subfield of E and E is a subfield of K, then $(K: F) = (K: E)(E: F)$.*

Note. The equation $(K: F) = (K: E)(E: F)$ is intended to include the statement that if either of the factors on the right is infinite then so is $(K: F)$.

Proof. Let $\{x_1, \ldots, x_r\}$ be a finite subset of E, linearly independent over F, and let $\{y_1, \ldots, y_s\}$ be a finite subset of K, linearly independent over E. We claim that the subset of K consisting of the rs elements $x_i y_j$ $(i = 1, \ldots, r;$ $j = 1, \ldots, s)$ is linearly independent over F. So suppose we have a relation

$$\sum_{i=1}^{r} \sum_{j=1}^{s} a_{ij} x_i y_j = 0,$$

with coefficients a_{ij} in F. We may write this relation in the form

$$\left(\sum_{i=1}^{r} a_{i1} x_i\right) y_1 + \ldots + \left(\sum_{i=1}^{r} a_{is} x_i\right) y_s = 0.$$

Since the coefficients $\sum_{i=1}^{r} a_{ij} x_i$ $(j = 1, \ldots, s)$ are elements of E and the set $\{y_1, \ldots, y_s\}$ is linearly independent over E it follows that these coefficients are all zero. Thus we have the s relations

$$a_{1j}x_1 + \ldots + a_{rj}x_r = 0 \quad (j = 1, \ldots, s).$$

The coefficients of each of these relations are elements of F, and hence, since the set $\{x_1, \ldots, x_r\}$ is linearly independent over F, we deduce that all the coefficients a_{ij} are zero ($i = 1, \ldots, r$; $j = 1, \ldots, s$). This is precisely what we had to show in order to prove that the set $\{x_i y_j\}$ is linearly independent over F.

As in Theorem 7.1, this completes the proof if either $(E: F)$ or $(K: E)$ is infinite. For then we may take r or s arbitrarily large and hence obtain arbitrarily large subsets of K linearly independent over F, so showing that $(K: F)$ is infinite.

Suppose, on the other hand, that $(E: F)$ and $(K: E)$ are both finite. Then we may take $r = (E: F)$ and $s = (K: E)$, and we may suppose that $\{x_1, \ldots, x_r\}$ and $\{y_1, \ldots, y_s\}$ are bases for E over F and K over E respectively (see Theorem 4.3). The preceding arguments, combined with Theorem 4.1, show that $(K: F) \geqq rs$. We shall now show that $\{x_i y_j\}$ is a generating system for K over F. It will follow from Theorem 4.2 that $(K: F) \leqq rs$ and the theorem will be established.

Let t, then, be any element of K. Since $\{y_1, \ldots, y_s\}$ is a generating system for K over E there exist elements $b_1, \ldots, b_s$ of E such that

$$t = b_1 y_1 + \ldots + b_s y_s.$$

Similarly, since $\{x_1, \ldots, x_r\}$ is a generating system for E over F, there are elements a_{ij} of F ($i = 1, \ldots, r$; $j = 1, \ldots, s$) such that

$$b_j = a_{1j}x_1 + \ldots + a_{rj}x_r \quad (j = 1, \ldots, s).$$

It follows at once that

$$t = \sum_{i=1}^{r} \sum_{j=1}^{s} a_{ij}x_i y_j;$$

that is to say, the set $\{x_i y_j\}$ generates K over F as required.
This completes the proof.

Theorem 7.2 has two useful corollaries. The first follows from the theorem by a simple inductive argument.

COROLLARY 1. *Let $F_0, F_1, \ldots, F_n$ be fields such that F_{i-1} is a subfield of F_i ($i = 1, \ldots, n$). Then*

$$(F_n: F_0) = (F_n: F_{n-1})(F_{n-1}: F_{n-2})\ldots(F_2: F_1)(F_1: F_0).$$

COROLLARY 2. *Let F be a subfield of K such that $(K: F)$ is finite. If E is a subfield of K including F such that $(E: F) = (K: F)$, then $E = K$.*

Proof. According to the theorem, $(K:F) = (K: E)(E:F)$. Hence $(K: E) = 1$, and the result follows from the remarks preceding the theorem.

We remark that if $(K: F)$ is infinite we cannot deduce from the equations $(K: F) = (K: E)(E: F)$ and $(E: F) = (K: F)$ that $(K: E) = 1$.

§ **8. Simple extensions.** Let E be a field, F a subfield of E and S any subset of E. We consider the intersection of all the subfields of E which include both the subfield F and all the elements of S – that is to say the subset of E consisting of the elements common to all those subfields. It is easy to verify that this intersection is itself a subfield of E including F and S; clearly it is the smallest such subfield. We call it the subfield of E **generated over** F **by** S and we denote it by $F(S)$; we also say that $F(S)$ is the subfield of E obtained by **adjoining** S to F. The next theorem gives us a description of the elements of $F(S)$.

THEOREM 8.1. *With the notation just described, the elements of $F(S)$ are precisely those elements of E which can be expressed as quotients of finite linear combinations (with coefficients in F) of finite products of elements of S.*

Proof. Let K be the subset of E consisting of elements which can be expressed in this way. Since the sum, difference, product and quotient of any two such elements can again be expressed in the same form, it follows that K is a subfield of E. Further, it is clear that K includes both F and S and hence includes $F(S)$.

On the other hand, since $F(S)$ is a subfield of E including F and S, we see at once that all finite products of elements of S, all finite linear combinations of such products with coefficients in F, and hence finally all quotients of such linear combinations belong to $F(S)$. That is to say, $F(S)$ includes K.

Thus $F(S) = K$, as asserted.

If the subset S consists of finitely many elements $\alpha_1, \ldots, \alpha_r$ of E, we usually write $F(\alpha_1, \ldots, \alpha_r)$ instead of $F(S)$. If, in particular, the subset S consists of a single element α of E, we call $F(S) = F(\alpha)$ a **simple extension** of F. (We recall our agreement that, when we refer to a field, such as $F(\alpha)$, as an extension of one of its subfields, the monomorphism of the extension is tacitly understood to be the inclusion monomorphism.) Thus a subfield K of E is a simple extension of the subfield F if there is an element α of K such that K is itself the smallest subfield of E which contains both F and α. It follows at once from Theorem 8.1 that $F(\alpha)$ consists of those elements of E which can be expressed as quotients of polynomials in α with coefficients in F. We now subject simple extensions to a more searching analysis.

THEOREM 8.2. *Let E be a field, F a subfield of E, α an element of E. Then either* (1) *$F(\alpha)$ is isomorphic to the field $R(F)$ of rational functions with coefficients in F, or* (2) *$F(\alpha)$ coincides with the ring $F[\alpha]$ of polynomials in α with coefficients in F. In the second case there is a unique monic irreducible polynomial m_α in $P(F)$ such that a polynomial f in $P(F)$ has α as a root if and only if it is a multiple of m_α. Further, $(F(\alpha): F) = \partial m_\alpha$.*

Proof. Let A_α be the set of polynomials in $P(F)$ which have α as a root; then a polynomial f in $P(F)$ belongs to A_α if and only if $\sigma_\alpha(f) = 0$, where σ_α is the substitution epimorphism of $P(F)$ onto $F[\alpha]$. Two cases can arise.

Case 1. A_α consists of the zero polynomial alone.

Then we may define a mapping ϕ of $R(F)$ into $F(\alpha)$ by setting

$$\phi(f/g) = \sigma_\alpha(f)(\sigma_\alpha(g))^{-1}$$

for every element f/g of $R(F)$; we remark that since g is not the zero polynomial in $P(F)$, $\sigma_\alpha(g)$ is not zero and hence has a multiplicative inverse. It is easy to verify that ϕ is an isomorphism of $R(F)$ onto $F(\alpha)$.

Case 2. A_α does not consist of the zero polynomial alone.

Let g be a non-zero polynomial of least degree in A_α. Thus $\sigma_\alpha(g) = 0$, but if f is a non-zero polynomial in $P(F)$ with $\partial f < \partial g$ then $\sigma_\alpha(f) \neq 0$. Let m_α be the monic polynomial obtained on dividing all the coefficients of g by the leading coefficient. Then of course $\sigma_\alpha(m_\alpha) = 0$ also.

We show first that m_α is irreducible. So suppose we have $m_\alpha = hk$, where h, k are polynomials in $P(F)$ with $0 < \partial h, \partial k < \partial m_\alpha$. Since σ_α is a homomorphism (Theorem 5.4) we have

$$0 = \sigma_\alpha(m_\alpha) = \sigma_\alpha(hk) = \sigma_\alpha(h)\sigma_\alpha(k).$$

Hence either $\sigma_\alpha(h) = 0$ or $\sigma_\alpha(k) = 0$; but this is impossible since m_α is a non-zero polynomial of least degree in A_α. It follows that m_α is irreducible, as asserted.

Next we claim that A_α consists precisely of the multiples of m_α. It is clear that all the multiples of m_α belong to A_α, for if h is any polynomial in $P(F)$,

$$\sigma_\alpha(m_\alpha h) = \sigma_\alpha(m_\alpha)\sigma_\alpha(h) = 0.$$

Suppose conversely that f is any polynomial in A_α, i.e., that $\sigma_\alpha(f) = 0$. According to Theorem 5.1 there exist polynomials q and r in $P(F)$ such that $f = qm_\alpha + r$ and $\partial r < \partial m_\alpha$. Then

$$0 = \sigma_\alpha(f) = \sigma_\alpha(q)\sigma_\alpha(m_\alpha) + \sigma_\alpha(r) = \sigma_\alpha(r).$$

Hence r is the zero polynomial and $f = qm_\alpha$; for otherwise we should have a contradiction to our choice of m_α as a non-zero polynomial of least degree in A_α.

It now follows at once that m_α is the unique monic irreducible polynomial in A_α. For if m'_α were another such polynomial the argument of the preceding paragraph applied to m_α and to m'_α would show that m_α is a factor of m'_α and m'_α is a factor of m_α; hence, since both polynomials are monic, $m_\alpha = m'_\alpha$.

We shall show next that $F(\alpha) = F[\alpha]$, i.e., that every element of $F(\alpha)$ can be expressed as a polynomial in α with coefficients in F. As we remarked just before the statement of this theorem, every element of $F(\alpha)$ has the form $\sigma_\alpha(f)(\sigma_\alpha(g))^{-1}$ where f and g are polynomials in $P(F)$ and, of course, $\sigma_\alpha(g)$ is non-zero. Since $\sigma_\alpha(g)$ is non-zero, g does not belong to A_α and hence is not divisible by m_α. Since m_α is irreducible, it follows that the highest common factor of g and m_α is the identity polynomial e. Then, according to Theorem 5.2, there exist polynomials h and k in $P(F)$ such that $gh + m_\alpha k = e$. Hence

$$e = \sigma_\alpha(e) = \sigma_\alpha(g)\sigma_\alpha(h) + \sigma_\alpha(m_\alpha)\sigma_\alpha(k) = \sigma_\alpha(g)\sigma_\alpha(h),$$

since $\sigma_\alpha(m_\alpha) = 0$; thus $(\sigma_\alpha(g))^{-1} = \sigma_\alpha(h)$ and so

$$\sigma_\alpha(f)(\sigma_\alpha(g))^{-1} = \sigma_\alpha(f)\sigma_\alpha(h) = \sigma_\alpha(fh),$$

which is a polynomial in α with coefficients in F, as required.

Finally, we claim that $(F(\alpha): F) = \partial m_\alpha$. We shall establish this result by showing that if $\partial m_\alpha = n$ then the set $\{e, \alpha, \alpha^2, ..., \alpha^{n-1}\}$ is a basis for $F(\alpha)$ when considered in the usual way as a vector space over F.

First we show that this set is linearly independent

over F; so suppose we have a relation of the form

$$a_0 e + a_1\alpha + \ldots + a_{n-1}\alpha^{n-1} = 0.$$

This relation may also be written

$$\sigma_\alpha(f) = 0,$$

where f is the polynomial $a_0 + a_1 X + \ldots + a_{n-1}X^{n-1}$. Since f belongs to A_α and $\partial f \leqq n-1 < \partial m_\alpha$ it follows by an argument which is by now familiar that f is the zero polynomial, i.e., that $a_0 = a_1 = \ldots = a_{n-1} = 0$. Thus $\{e, \alpha, \alpha^2, \ldots, \alpha^{n-1}\}$ is linearly independent over F.

Next we show that $\{e, \alpha, \alpha^2, \ldots, \alpha^{n-1}\}$ is a generating system for $F(\alpha)$ over F. So let x be any element of $F(\alpha)$; as we have just seen, there exists a polynomial f in $P(F)$ such that $x = \sigma_\alpha(f)$. Since we can express f in the form $f = qm_\alpha + r$, with $\partial r < n$, it follows that $x = \sigma_\alpha(f) = \sigma_\alpha(r)$, which is a linear combination of $e, \alpha, \alpha^2, \ldots, \alpha^{n-1}$ with coefficients in F.

This completes the proof that $\{e, \alpha, \alpha^2, \ldots, \alpha^{n-1}\}$ is a basis for $F(\alpha)$ over F and hence that $(F(\alpha){:}\ F) = n$.

In Case 1 of Theorem 8.2, where A_α consists of the zero polynomial alone, we say that the element α of E is **transcendental over** F. Thus if α is transcendental over F a polynomial in α with coefficients in F can be zero only if all the coefficients are zero. In Case 2 of the theorem, where A_α does not consist of the zero polynomial alone, α is said to be **algebraic over** F; in this case the unique monic irreducible polynomial m_α in A_α is called the **minimum polynomial** of α over F. We may denote this polynomial by $m_{\alpha, F}$ when we wish to bring the field F into prominence. According to the theorem, a polynomial in $P(F)$ has α as a root if and only if it is a multiple of the minimum polynomial of α over F.

In the following examples we shall assume that the field **Q** of rational numbers is identified with a subfield of the field **R** of real numbers, that **R** is in turn identified with a subfield of the field **C** of complex numbers and that all the monomorphisms involved in the extensions considered are the appropriate inclusion monomorphisms.

Example 1. For this one example let e denote the base of natural logarithms. Hermite proved in 1873 that e is not a root of any non-zero polynomial with rational number coefficients, i.e., that e is transcendental over the rational number field **Q**. It follows from the first part of Theorem 8.2 that the subfield $\mathbf{Q}(e)$ of the real number field **R** is isomorphic to the field of rational functions $R(\mathbf{Q})$.

Example 2. The complex number i is a root of the polynomial X^2+1 in $P(\mathbf{Q})$; so i is algebraic over **Q**. Since X^2+1 is in fact irreducible in $P(\mathbf{Q})$ it follows from the second part of Theorem 8.2 that the subfield $\mathbf{Q}(i)$ of the complex number field **C** has degree 2 over **Q**; X^2+1 is the minimum polynomial of i over **Q**. Every element of $\mathbf{Q}(i)$ can be expressed uniquely in the form $a+bi$ where a and b are rational numbers.

§ 9. Algebraic extensions. Let now $(E, \imath)$ be any extension of a field F. We say that $(E, \imath)$ is an **algebraic extension** of F if every element α of E is algebraic over the subfield $\imath(F)$. Otherwise $(E, \imath)$ is called a **transcendental extension** of F. As the following theorem shows, this classification of extensions into algebraic and transcendental is not unconnected with our earlier classification into finite and infinite extensions.

THEOREM 9.1. *Every finite extension of a field is algebraic.*

Proof. Let $(E, \imath)$ be a finite extension of a field F; suppose $(E: \imath(F)) = n$.

Let α be any element of E. We claim that the set $A = \{\imath(e), \alpha, \alpha^2, \ldots, \alpha^n\}$ is linearly dependent over $\imath(F)$. This is clear if any two elements are equal; otherwise A consists of $n+1$ elements of the n-dimensional vector space

E and so is linearly dependent. That is to say, there exist elements $a_0, \alpha_1, \ldots, a_n$ of F, not all zero, such that

$$\imath(a_0)\imath(e)+\imath(a_1)\alpha+\imath(a_2)\alpha^2+\ldots+\imath(a_n)\alpha^n = 0.$$

Thus α is algebraic over $\imath(F)$ as required.

The converse of this theorem is not true: it is possible to construct algebraic extensions which are not finite. There is, however, a partial converse, as follows.

THEOREM 9.2. *Let $(E, \imath)$ be an extension of a field F. If E is generated over $\imath(F)$ by a finite set of elements algebraic over $\imath(F)$, then $(E, \imath)$ is a finite extension of F.*

Proof. We shall identify F and $\imath(F)$ and suppose that $E = F(\alpha_1, \ldots, \alpha_n)$ where $\alpha_1, \ldots, \alpha_n$ are algebraic over F.

Set $E_0 = F$ and $E_k = F(\alpha_1, \ldots, \alpha_k)$ $(k = 1, 2, \ldots, n)$, so that $E_n = E$. Then for $k = 1, 2, \ldots, n$, E_k is a simple extension of E_{k-1}: $E_k = E_{k-1}(\alpha_k)$. Since α_k is algebraic over F it is *a fortiori* algebraic over E_{k-1} and hence, according to Theorem 8.2, $(E_k\colon E_{k-1})$ is finite, being equal to the degree of the minimum polynomial of α_k over E_{k-1}.

Thus $(E\colon F) = \prod_{k=1}^{n} (E_k\colon E_{k-1})$ is also finite; that is to say, $(E, \imath)$ is a finite extension of F.

Let F be a subfield of a field E. Then the **relative algebraic closure** of F in E is the subset of E consisting of all those elements of E which are algebraic over F. If the relative algebraic closure of F in E is just the field F itself (which means that all the elements of E which do not belong to F are transcendental over F) we say that F is **relatively algebraically closed** in E.

THEOREM 9.3. *If F is a subfield of a field E, the relative algebraic closure A of F is a subfield of E which is algebraic over F and includes all the subfields of E which are algebraic over F.*

Proof. Apart from the assertion that A is a subfield of E, this theorem is an immediate consequence of the definition.

To show that A is a subfield we have only to prove that if α and β are any two elements of E which are algebraic over F then $\alpha+\beta$, $\alpha\beta$, $-\alpha$ and (when α is non-zero) α^{-1} are also algebraic over F. These elements, however, all belong to the subfield $F(\alpha, \beta)$ of E obtained by adjoining $\{\alpha, \beta\}$ to F. It follows from Theorem 9.2 that $F(\alpha, \beta)$ is a finite extension of F and hence, from Theorem 9.1, that $F(\alpha, \beta)$ is an algebraic extension of F.

This completes the proof.

Example. Suppose that the rational number field $\mathbf{Q}$ has been identified with a subfield of the complex number field $\mathbf{C}$. The relative algebraic closure $\mathbf{A}$ of $\mathbf{Q}$ in $\mathbf{C}$ is called the **field of algebraic numbers.** It can be shown that $\mathbf{A}$ is an infinite extension of $\mathbf{Q}$, thus providing us with an example of an algebraic extension which is not of finite degree. The field $\mathbf{A}$, however, does not coincide with the whole complex number field $\mathbf{C}$ since (for example) the base of natural logarithms does not belong to $\mathbf{A}$ (see § 8, Example 1).

§ 10. Factorisation of polynomials. Let (E, ι) be an extension of a field F. If κ_1 and κ_2 are the canonical monomorphisms of F into $P(F)$ and E into $P(E)$ respectively, it follows from Theorem 5.6 that there exists a monomorphism ι_P of $P(F)$ into $P(E)$ such that $\iota_P(\kappa_1(a)) = \kappa_2(\iota(a))$ for all elements a of F. Since we usually agree to identify F with $\kappa_1(F)$, E with $\kappa_2(E)$ and F with $\iota(F)$, it is natural that we should also agree to identify $P(F)$ with its image under ι_P. When we make this identification what we are doing in effect is to regard polynomials with coefficients in F also as polynomials with coefficients in E.

Let E be a field, F a subfield of E, α an element of E

algebraic over F. Since α is a root in $F(\alpha)$ of its minimum polynomial $m_{\alpha, F}$ (considered as a polynomial in $P(F(\alpha))$), it follows from Theorem 5.5 that the linear polynomial $l_\alpha = X-\alpha$ is a factor of $m_{\alpha, F}$ in $P(F(\alpha))$. Thus although $m_{\alpha, F}$ is irreducible in $P(F)$ it has a linear factor when we consider it as a polynomial in $P(F(\alpha))$. We now naturally ask whether, given an irreducible polynomial f in $P(F)$, it is possible to find an extension (E, ι) of F such that $\iota_P(f)$ has a linear factor in $P(E)$. The next theorem answers this question in the affirmative.

THEOREM 10.1. *Let F be a field, f a non-constant irreducible polynomial in P(F). Then there exists an extension (E, ι) of F such that $\iota_P(f)$ has a linear factor l_α in P(E). Further, the degree of $\iota(F)(\alpha)$ over F is precisely ∂f.*

Proof. We define a relation between the polynomials in $P(F)$ by saying that a polynomial g is congruent to a polynomial h modulo f if the difference $g-h$ is divisible in $P(F)$ by the given polynomial f. The set of all polynomials in $P(F)$ congruent to g is called the residue class of g modulo f; it is clearly non-empty, since g itself belongs to it. Let E be the set of residue classes so obtained.

We now introduce operations of addition and multiplication in E. If C_1, C_2 are residue classes in E, we select polynomials f_1, f_2 from C_1, C_2 respectively and we define C_1+C_2 and C_1C_2 to be the residue classes of f_1+f_2 and f_1f_2 respectively. A rather long, but entirely routine, investigation shows that these residue classes depend only on C_1 and C_2 and not on the choice of representative polynomials f_1, f_2, and that under the operations of addition and multiplication so defined E forms a commutative ring. The zero element Z and the identity element I are the residue classes modulo f of the zero polynomial and the identity polynomial respectively. (Cf. Example 4 of § 1.)

We show next that E is actually a field. So let C_1 be any non-zero residue class, f_1 any polynomial in C_1. Then f_1 is not congruent to the zero polynomial modulo f, and hence is not divisible by f. Since f is irreducible it follows that the highest common factor of f_1 and f is the identity polynomial e. Thus, according to Theorem 5.2, there exist polynomials a and b in $P(F)$ such that $af+bf_1 = e$. Then $bf_1-e = -af$ is divisible by f and consequently bf_1 is congruent modulo f to the identity polynomial. Hence, if C_2 is the residue class of the polynomial b, we have $C_2C_1 = I$. Thus we have shown that every non-zero element of E has an inverse relative to the multiplication in E; so E is a field.

Now consider the mapping $\imath$ of F into E obtained by defining $\imath(a)$ for each element a of F to be the residue class modulo f of the constant polynomial a. It is easy to verify that $\imath$ is a homomorphism. Since $\imath(e) = I$ and I is not the zero element of E it follows that $\imath$ is not the zero homomorphism and hence, by Theorem 3.1, is a monomorphism of F into E. Thus $(E, \imath)$ is an extension of F.

Let α be the residue class of the polynomial X. Then E is a simple extension obtained by adjoining α to $\imath(F)$. To see this, let C be any residue class modulo f, and let $g = b_0+b_1X+\ldots+b_rX^r$ be any polynomial in C; then we have

$$C = \imath(b_0)+\imath(b_1)\alpha+\ldots+\imath(b_r)\alpha^r = \sigma_\alpha(\imath_P(g)).$$

In particular, since the residue class containing the given irreducible polynomial f is the zero class Z, we have $\sigma_\alpha(\imath_P(f)) = Z$. Thus $\imath_P(f)$ has α as a root in E and l_α as a linear factor in $P(E)$. Since f is irreducible in $P(F)$ $\imath_P(f)$ is irreducible in $P(\imath(F))$ and hence $\imath_P(f)$ differs from the minimum polynomial of α over $\imath(F)$ only by a constant factor. It follows from Theorem 8.2 that the degree of $(E, \imath)$ over F is exactly $\partial(\imath_P(f)) = \partial f$.

Although we have given only one method of constructing an extension in which a given irreducible polynomial has a root, it will follow as a special case of the next theorem that any other method of constructing such an extension will lead to essentially the same result.

THEOREM 10.2. *Let F and F' be fields, τ an isomorphism of F onto F' and τ_P the canonical extension of τ to $P(F)$. Let f be an irreducible polynomial in $P(F)$ and let (E, ι) and (E', ι') be extensions of F and F' in which f and $\tau_P(f)$ have roots β and β' respectively. Let K and K' be the subfields of E and E' generated over $\iota(F)$ and $\iota'(F')$ by β and β' respectively. Then there exists an isomorphism τ_1 of K onto K' such that $\tau_1(\beta) = \beta'$ and $\tau_1(\iota(a)) = \iota'(\tau(a))$ for every element a of F.*

Proof. Let $\partial f = r$. According to Theorem 8.2, every element of K can be expressed uniquely in the form $\sum_{k=0}^{r-1} \iota(a_k)\beta^k$ where $a_0, \ldots, a_{r-1}$ are elements of F; similarly, every element of K' can be expressed uniquely in the form $\sum_{k=0}^{r-1} \iota'(a'_k)(\beta')^k$ where $a'_0, \ldots, a'_{r-1}$ are elements of F'. The mapping τ_1 of K into K' defined by setting

$$\tau_1\left(\sum_{k=0}^{r-1} \iota(a_k)\beta^k\right) = \sum_{k=0}^{r-1} \iota'(\tau(a_k))(\beta')^k$$

is easily seen to satisfy the requirements of the theorem.

The assertion which we made just before the statement of the theorem follows at once if we take $F = F'$, $\tau = \varepsilon$, the identity automorphism of F, and recall that isomorphic fields K and K' may be regarded as "essentially the same".

We sometimes put the conclusion of the theorem in pictorial terms by saying that it is possible to insert an arrow from K to K' in the diagram in fig. 2 so that the diagram becomes commutative. (Cf. Theorem 5.6.) If we identify F and F' with their images under ι and ι' respectively, the condition on τ_1 simply reduces to the requirement that $\tau_1(a) = \tau(a)$, i.e., that τ_1 shall act like τ on the elements of F. We call τ_1 an **extension** of τ.

In the following examples we shall assume, as in § 8, that **Q** is identified with a subfield of **R**, that **R** is identified with a subfield of **C** and that all the monomorphisms involved in the extensions considered are the appropriate inclusion monomorphisms.

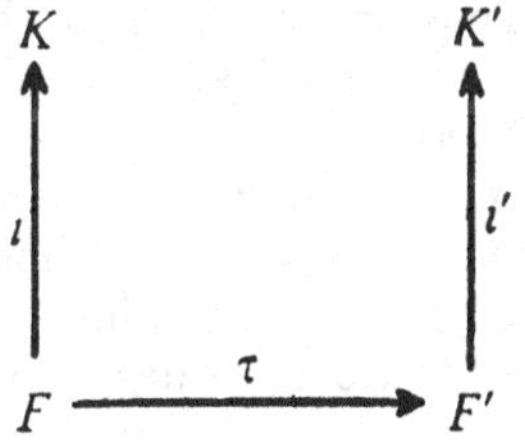

$$\begin{array}{ccc} K & & K' \\ \uparrow \iota & & \uparrow \iota' \\ F & \xrightarrow{\tau} & F' \end{array}$$

FIG. 2

Example 1. Let f be the polynomial X^3-2 in $P(\mathbf{Q})$. We claim that f is irreducible in $P(\mathbf{Q})$. For, if it were reducible in $P(\mathbf{Q})$, at least one of its monic irreducible factors in $P(\mathbf{Q})$ would be linear, say $X-\alpha$; set $\alpha = p/q$ where p and q are integers which have no common factor. Then α is a root of X^3-2; so we have $p^3/q^3-2 = 0$, whence $p^3/q = 2q^2$, which is impossible unless $q = 1$, since if $q \neq 1$ the right hand member is an integer while the left hand member is not. Hence, if X^3-2 is reducible in $P(\mathbf{Q})$ it has an integral root. But this is impossible, since there is no integer whose cube is 2. So X^3-2 is irreducible in $P(\mathbf{Q})$.

The mapping ϕ of **R** into **R** defined by setting $\phi(x) = x^3$ for all real numbers x is a continuous function; since $\phi(0) = 0$ and $\phi(2) = 8$, it follows from a well-known result on continuous functions that there exists a positive real number α such that $\phi(\alpha) = \alpha^3 = 2$. Since ϕ is an increasing function, α is the only positive real number with this property.

Consider now the polynomial $f = X^3-2$ as a polynomial with coefficients in $\mathbf{Q}(\alpha)$: we have the decomposition

$$f = X^3-\alpha^3 = (X-\alpha)(X^2+\alpha X+\alpha^2).$$

Next, in the ring $P(\mathbf{C})$ the quadratic factor decomposes further:

$$X^2+\alpha X+\alpha^2 = (X-\tfrac{1}{2}(-1+i\sqrt{3})\alpha)(X-\tfrac{1}{2}(-1-i\sqrt{3})\alpha).$$

Thus in the field $\mathbf{C}$ the polynomial f has three roots, $\alpha, \beta = \frac{1}{2}(-1+i\sqrt{3})\alpha,\ \gamma = \frac{1}{2}(-1-i\sqrt{3})\alpha$. According to Theorem 10.2 there is an isomorphism τ of $\mathbf{Q}(\alpha)$ onto $\mathbf{Q}(\beta)$ such that $\tau(a) = a$ for all rational numbers a and $\tau(\alpha) = \beta$; and indeed all three fields $\mathbf{Q}(\alpha)$, $\mathbf{Q}(\beta)$, $\mathbf{Q}(\gamma)$ are isomorphic. Since f is irreducible in $P(\mathbf{Q})$ these fields are all of degree 3 over $\mathbf{Q}$.

Example 2. The polynomial X^2+1 is irreducible in $P(\mathbf{Q})$ and has two roots, i and $-i$, in $\mathbf{C}$. Hence, according to Theorem 10.2, there is an isomorphism τ of $\mathbf{Q}(i)$ onto $\mathbf{Q}(-i)$ such that $\tau(a) = a$ for all rational numbers a and $\tau(i) = -i$; this is given by $\tau(a+bi) = a-bi$ for all elements $a+bi$ of $\mathbf{Q}(i)$. The fields $\mathbf{Q}(i)$ and $\mathbf{Q}(-i)$ are in fact the same field; so τ is actually an automorphism of $\mathbf{Q}(i)$.

Theorem 10.1 assures us that it is always possible to construct an extension of a field F in which a given irreducible polynomial f of $P(F)$ has a root. Before we can drop the word " irreducible " from this assertion we must interpolate the following theorem which allows us to say that every polynomial has an irreducible factor.

THEOREM 10.3. *Let F be a field and let f be a non-constant polynomial in $P(F)$. Then there exist an element a of F and non-constant monic irreducible polynomials $f_1, f_2, \ldots, f_r$ in $P(F)$ such that $f = af_1f_2\ldots f_r$.*

Note. To be quite precise we ought to write $f = \kappa(a)f_1f_2\ldots f_r$ where κ is the canonical monomorphism of F into $P(F)$.

Proof. We proceed by induction on the degree n of the polynomial f.

If $n = 1$, so that $f = aX+b$, where a and b are elements of F and a is non-zero, we may write $f = a(X+a^{-1}b)$. Since $X+a^{-1}b$ is irreducible, the desired result is established in this case.

Suppose now that we have established the result for all polynomials of degree less than k. Let f be a polynomial of degree k. If f is itself irreducible, with leading coefficient a, let f_1 be the monic polynomial obtained on dividing all the coefficients of f by a. Then f_1 is irreducible and $f = af_1$ is an expression of the required form. On the other hand, if f is reducible, there exist polynomials g and h such that $f = gh$ and $\partial g, \partial h < k$. According to the inductive hypothesis there exist elements b, c of F and non-constant monic irreducible polynomials $g_1, g_2, \ldots, g_s$ and $h_1, h_2, \ldots, h_t$ such that $g = bg_1 g_2 \ldots g_s$ and $h = ch_1 h_2 \ldots h_t$. Thus $f = (bc)g_1 \ldots g_s h_1 \ldots h_t$, which is again an expression of the desired form.

This completes the induction.

Our next theorem includes as a special case the result that the monic irreducible factors $f_1, \ldots, f_r$ of f, whose existence is assured by Theorem 10.3, are uniquely determined by f apart from their order.

THEOREM 10.4. *Let F be a field and let f and g be monic polynomials in $P(F)$ such that g is a factor of f. If $f = f_1 \ldots f_r$, where $f_1, \ldots, f_r$ are non-constant monic irreducible polynomials in $P(F)$, then every non-constant monic irreducible factor of g in $P(F)$ belongs to the set $\{f_1, \ldots, f_r\}$.*

Proof. Suppose g_1 is a non-constant monic irreducible factor of g in $P(F)$; say $g = g_1 h_1$. Then, if $f = gh$, we have $f = g_1 h_1 h$.

Let us suppose that g_1 does not belong to the set $\{f_1, \ldots, f_r\}$. Then the highest common factor of g_1 and each of the polynomials $f_1, \ldots, f_r$ is the identity polynomial e. Hence, according to Theorem 5.2, there exist pairs of polynomials a_k, b_k in $P(F)$ such that

$$a_k g_1 + b_k f_k = e \quad (k = 1, 2, \ldots, r). \tag{10.1}$$

Let (E, ι) be an extension of F in which $\iota_P(g_1)$ has a root α; as usual we shall identify $P(F)$ and its image under ι_P in $P(E)$. Applying the substitution epimorphism σ_α to each of the equations (10.1), we obtain

$$\begin{aligned} e = \sigma_\alpha(e) &= \sigma_\alpha(a_k)\sigma_\alpha(g_1) + \sigma_\alpha(b_k)\sigma_\alpha(f_k) \\ &= \sigma_\alpha(b_k)\sigma_\alpha(f_k) \quad (k = 1, 2, \ldots, r). \end{aligned}$$

It follows that $\sigma_\alpha(f_k)$ is non-zero $(k = 1, 2, \ldots, r)$ and hence that $\sigma_\alpha(f) = \sigma_\alpha(f_1)\ldots\sigma_\alpha(f_r)$ is non-zero. But since $f = g_1 h_1 h$, we have $\sigma_\alpha(f) = \sigma_\alpha(g_1)\sigma_\alpha(h_1 h) = 0$.

So we have reached a contradiction, and hence g_1 belongs to the set $\{f_1, \ldots, f_r\}$ as asserted.

If we take $g = f$ in Theorem 10.4 we obtain the uniqueness result mentioned before the statement of the theorem.

We are now in a position to drop the hypothesis of irreducibility in Theorem 10.1.

THEOREM 10.5. *Let F be a field, f any non-constant polynomial in $P(F)$, not necessarily irreducible. Then there exists an extension (E, ι) of F such that $\iota_P(f)$ has a linear factor l_α in $P(E)$. Further, the degree of $\iota(F)(\alpha)$ over F is at most ∂f.*

Proof. If f is irreducible in $P(F)$ this is simply a restatement of Theorem 10.1; in this case the degree of $\iota(F)(\alpha)$ over F is exactly ∂f.

If, however, f is reducible in $P(F)$, let f_1 be a non-constant irreducible factor of f in $P(F)$. Then, according to Theorem 10.1, there exists an extension (E, ι) of F such that $\iota_P(f_1)$, and hence of course $\iota_P(f)$ also, has a linear factor l_α in $P(E)$. In this case the degree of $\iota(F)(\alpha)$ over F is ∂f_1 which is less than ∂f.

§ **11. Splitting fields.** Let K be a field, p a polynomial in $P(K)$. We say that the polynomial p **splits completely** in $P(K)$ if all its non-constant irreducible factors in $P(K)$ are linear, i.e., if there exist elements α, α_1, ..., α_k of K such that

$$p = \alpha(X-\alpha_1)(X-\alpha_2)\dots(X-\alpha_k) = \alpha l_{\alpha_1} l_{\alpha_2} \dots l_{\alpha_k}.$$

Now let F be a field, f a polynomial in $P(F)$. An extension (K, ι) of F is called a **splitting field for** f **over** F if (1) $\iota_P(f)$ splits completely in $P(K)$ and (2) $\iota_P(f)$ does not split completely in $P(E)$ where E is any subfield of K including $\iota(F)$ other than K itself. Clearly K is generated over $\iota(F)$ by the roots of $\iota_P(f)$ in K.

In Theorems 10.1 and 10.5 we saw how to construct an extension of a field F in which a given polynomial in $P(F)$ splits off one linear factor; our next theorem shows how, by repeated application of this procedure, we may construct an extension in which the polynomial splits completely into linear factors.

THEOREM 11.1. *Let F be a field, f a non-constant polynomial of degree n in $P(F)$. Then there exists a splitting field (K, ι) for f over F, and the degree of K over F is at most $n!$.*

Proof. According to Theorem 10.5 there exists an extension (E_1, ι_1) of F such that $(\iota_1)_P(f)$ has at least one linear factor in $P(E_1)$ and the degree of E_1 over F is at most n.

We now proceed inductively. Let r be any integer, $1 \leqq r < n$, and suppose we have constructed an extension (E_r, ι_r) of F such that $(\iota_r)_P(f)$ has at least r linear factors in $P(E_r)$ and the degree of E_r over F is at most

$$n(n-1)\dots(n-r+1).$$

If $(\iota_r)_P(f) = l_{\beta_1} \dots l_{\beta_r} f'$ in $P(E_r)$, we have $\partial f' = n-r$, and so, according to Theorem 10.5, there is an extension (E_{r+1}, ι') of E_r such that $\iota'_P(f')$ has at least one linear factor in $P(E_{r+1})$ and the degree of E_{r+1} over E_r is at most $n-r$. Let ι_{r+1} be the mapping of F into E_{r+1} defined by setting $\iota_{r+1}(a) = \iota'(\iota_r(a))$ for every element a of F. Then it is clear that (E_{r+1}, ι_{r+1}) is an extension of F such that $(\iota_{r+1})_P(f)$ has at least $r+1$ linear factors in $P(E_{r+1})$ and the degree of E_{r+1} over F is at most $n(n-1)...(n-r)$.

Thus after n steps we obtain an extension $(E, \iota) = (E_n, \iota_n)$ of F such that $\iota_P(f)$ has n linear factors $-l_{\alpha_1}, ..., l_{\alpha_n}$ say —in $P(E)$ and the degree of E over F is at most $n!$. Let K be the subfield of E obtained by adjoining $\alpha_1, ..., \alpha_n$ to $\iota(F)$. Then (K, ι) is a splitting field for f over F and the degree of K over F is at most $n!$.

If (K, ι) is a splitting field over F for a polynomial f in $P(F)$ we may deduce from Theorem 10.4 that the monic linear factors of $\iota_P(f)$ in $P(K)$ are unique apart from their order, and also that, if g is any factor of f in $P(F)$, $\iota_P(g)$ splits completely in $P(K)$ and its monic linear factors (and hence its roots) are among those of $\iota_P(f)$.

We now prove an analogue of Theorem 10.2 for splitting fields, which establishes in effect that all splitting fields of a polynomial are isomorphic.

THEOREM 11.2. *Let F and F' be fields, τ an isomorphism of F onto F' and τ_P the canonical extension of τ to $P(F)$. Let f be a polynomial in $P(F)$. If (K, ι) and (K', ι') are splitting fields for f and $\tau_P(f)$ over F and F' respectively, then there exists an isomorphism τ_1 of K onto K' such that $\tau_1(\iota(a)) = \iota'(\tau(a))$ for every element a of F.*

Proof. We shall identify F and F' with their images $\iota(F)$ and $\iota'(F')$ in K and K' respectively. The condition on the required isomorphism τ_1 then simply reduces to the requirement that τ_1 shall act like the given isomorphism τ on the elements of F; as usual in this situation we shall call τ_1 an extension of τ.

Suppose that in $P(K)$ we have the factorisation

$$f = \alpha(X-\alpha_1)(X-\alpha_2)\ldots(X-\alpha_r). \qquad (11.1)$$

We shall proceed by induction on the number n of roots α_i of f in K which do not belong to F.

If $n = 0$, then all the roots α_i belong to F and hence the splitting (11.1) actually occurs in $P(F)$ itself: so F is itself a splitting field for f over F, i.e., $K = F$. In this case when we apply the isomorphism τ_P to the equation (11.1) we obtain

$$\tau_P(f) = \tau(\alpha)(X-\tau(\alpha_1))(X-\tau(\alpha_2))\ldots(X-\tau(\alpha_r)).$$

Hence $\tau_P(f)$ splits completely in $P(F')$: so F' is a splitting field for $\tau_P(f)$ over F', i.e., $K' = F'$. Thus in this case τ is itself the required extension of τ to the splitting field.

Next suppose that there are $k>0$ roots α_i of f in K but not in F. We make the inductive hypothesis that, for every polynomial with coefficients in a field E which has fewer than k roots outside E in a splitting field L including E, we can extend every isomorphism of E to an isomorphism of the splitting field L.

The irreducible factors of f in $P(F)$ are not all linear—otherwise all the roots of f would belong to F and we are supposing that this is not the case. So let f_1 be a non-linear irreducible factor of f in $P(F)$. Then $\tau_P(f_1)$ is an irreducible factor of $\tau_P(f)$ in $P(F')$. As we remarked above (just before the statement of the present theorem) the roots of f_1 in K are included among the roots α_i of f in K: say α_1 is a root of f_1 in K. Similarly one of the roots of $\tau_P(f)$ in K', say β_1, is a root of $\tau_P(f_1)$ in K'. According to Theorem 10.2 there exists an isomorphism τ' of $F(\alpha_1)$ onto $F'(\beta_1)$ which acts like τ on F.

Now K and K' are clearly splitting fields for f and $\tau_P(f)$ over $F(\alpha_1)$ and $F'(\beta_1)$ respectively; but f has fewer than k roots in K outside $F(\alpha_1)$. Hence, by the inductive hypothesis, there exists an isomorphism τ_1 of K onto K' which acts like τ' on $F(\alpha_1)$ and hence like τ on F.

This completes the induction.

The various fields involved in the proof of this theorem and the mappings between them are illustrated in the diagram of fig. 3, where the unlettered arrows denote inclusion monomorphisms and the diagram is commutative in the sense described after Theorem 5.6.

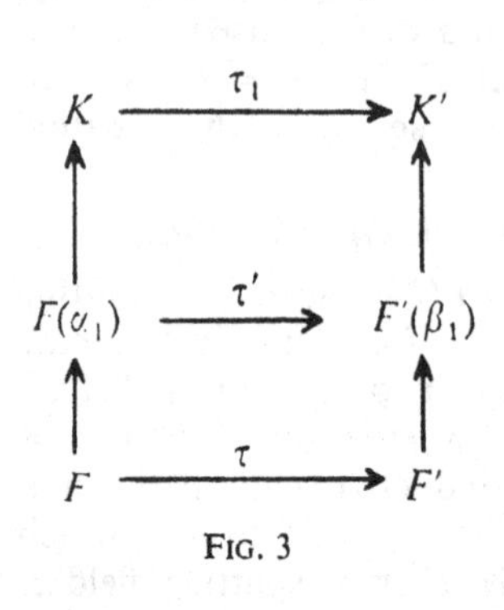

FIG. 3

We now give some examples of the construction of splitting fields for polynomials with rational number coefficients. As usual, we shall assume that the field **Q** of rational numbers is identified with a subfield of the field **R** of real numbers, that **R** is in turn identified with a subfield of the field **C** of complex numbers and that all the monomorphisms of the extensions considered are the appropriate inclusion monomorphisms.

Example 1. X^2-2X+1.

Since $X^2-2X+1 = (X-1)^2$ already splits completely in $P(\mathbf{Q})$ it follows that **Q** is itself a splitting field over **Q** for this polynomial.

Example 2. X^2-X+1.

This polynomial splits completely in the polynomial ring $P(\mathbf{C})$:

$$X^2-X+1 = (X-\tfrac{1}{2}(1+i\sqrt{3}))(X-\tfrac{1}{2}(1-i\sqrt{3})).$$

To obtain a splitting field, however, we need only take the subfield of **C** obtained by adjoining to **Q** the two complex numbers $\frac{1}{2}(1+i\sqrt{3})$ and $\frac{1}{2}(1-i\sqrt{3})$. But this subfield may be obtained equally well by adjoining the single complex number $i\sqrt{3}$; so $\mathbf{Q}(i\sqrt{3})$ is a splitting field for X^2-X+1 over $\mathbf{Q}$. First of all we know that $i\sqrt{3}$ is not a rational number, since its square is negative; so $\mathbf{Q}(i\sqrt{3}) \neq \mathbf{Q}$, and hence $(\mathbf{Q}(i\sqrt{3})\colon \mathbf{Q})>1$. Next $i\sqrt{3}$ is a root of the second-degree polynomial X^2+3 in $P(\mathbf{Q})$; hence $(\mathbf{Q}(i\sqrt{3})\colon \mathbf{Q})\leq 2$. It follows at once that

$$(\mathbf{Q}(i\sqrt{3})\colon \mathbf{Q}) = 2.$$

Example 3. X^3-2.

We have already seen in Example 1 of § 10 that there is a unique positive real number α such that $\alpha^3 = 2$, that $(\mathbf{Q}(\alpha)\colon \mathbf{Q}) = 3$ and that in $P(\mathbf{C})$ we have the decomposition

$$X^3-2 = (X-\alpha)(X-\tfrac{1}{2}(-1+i\sqrt{3})\alpha)(X-\tfrac{1}{2}(-1-i\sqrt{3})\alpha).$$

From these remarks it follows easily that $K = \mathbf{Q}(\alpha,\, i\sqrt{3})$ is a splitting field for X^3-2 over $\mathbf{Q}$.

We proceed to determine its degree over $\mathbf{Q}$. Since $i\sqrt{3}$ is a root of the second-degree polynomial X^2+3 in $P(\mathbf{Q}(\alpha))$ it follows that $(K\colon \mathbf{Q}(\alpha))\leq 2$. If we can show that $i\sqrt{3}$ does not belong to $\mathbf{Q}(\alpha)$ it will follow that $(K\colon \mathbf{Q}(\alpha)) = 2$. Suppose, to the contrary, that $i\sqrt{3}$ does lie in $\mathbf{Q}(\alpha)$; then, according to Theorem 8.2, there are rational numbers a, b, c such that $i\sqrt{3} = a+b\alpha+c\alpha^2$. Squaring this equation we obtain

$$\begin{aligned} -3 &= a^2+b^2\alpha^2+c^2\alpha^4+2ab\alpha+2bc\alpha^3+2ca\alpha^2 \\ &= (a^2+4bc)+(2c^2+2ab)\alpha+(b^2+2ca)\alpha^2. \end{aligned}$$

Since $\{1,\, \alpha,\, \alpha^2\}$ is a basis for $\mathbf{Q}(\alpha)$ over $\mathbf{Q}$ (Theorem 8.2), it follows that

$$a^2+4bc = -3,\; c^2+ab = 0,\; b^2+2ca = 0.$$

Neither b nor c is zero, since otherwise we should have $a^2 = -3$, which is impossible. Then $ab = -c^2$ is negative, so a and b have opposite sign; similarly a and c have opposite sign. Thus b and c have the same sign. But this implies that $-3 = a^2+4bc$ is positive, and this is a contradiction. Hence $(K: \mathbf{Q}(\alpha)) = 2$ and so

$$(K: \mathbf{Q}) = (K: \mathbf{Q}(\alpha))(\mathbf{Q}(\alpha): \mathbf{Q}) = 6.$$

Example 4. X^4-2.

As in Example 1 of § 10 we can show that there is a unique positive real number α such that $\alpha^4 = 2$; so we begin by considering the subfield $\mathbf{Q}(\alpha)$ of $\mathbf{R}$. We claim that $\mathbf{Q}(\alpha)$ has degree 4 over $\mathbf{Q}$. To this end we remark first that α^2 is a root of the polynomial X^2-2, which is irreducible in $P(\mathbf{Q})$ (the irreducibility follows from an argument similar to that for X^3-2 in Example 1 of § 10). Hence

$$(\mathbf{Q}(\alpha^2): \mathbf{Q}) = 2,$$

and $\{1, \alpha^2\}$ is a basis for $\mathbf{Q}(\alpha^2)$ over $\mathbf{Q}$. Next, α is a root of the second-degree polynomial $X^2-\alpha^2$ in $P(\mathbf{Q}(\alpha^2))$; so $(\mathbf{Q}(\alpha): \mathbf{Q}(\alpha^2)) \leqq 2$. If we can show that α does not belong to $\mathbf{Q}(\alpha^2)$ it will follow that $(\mathbf{Q}(\alpha): \mathbf{Q}(\alpha^2)) = 2$. So suppose, to the contrary, that α belongs to $\mathbf{Q}(\alpha^2)$. Then there exist rational numbers a and b such that $\alpha = a+b\alpha^2$, whence $\alpha^2 = a^2+2ab\alpha^2+b^2\alpha^4 = a^2+2b^2+2ab\alpha^2$. It follows that $a^2+2b^2 = 0$ (whence $a = b = 0$) and $2ab = 1$; this is, of course, a contradiction. So $(\mathbf{Q}(\alpha): \mathbf{Q}(\alpha^2)) = 2$ and hence $(\mathbf{Q}(\alpha): \mathbf{Q}) = 4$.

Consider now the polynomial X^4-2 as a polynomial with coefficients in $\mathbf{Q}(\alpha)$: we have the decomposition

$$X^4-2 = X^4-\alpha^4 = (X-\alpha)(X+\alpha)(X^2+\alpha^2).$$

Next, in the ring $P(\mathbf{C})$, the quadratic factor splits completely:

$$X^2+\alpha^2 = (X-\alpha i)(X+\alpha i).$$

It now follows easily that $K = \mathbf{Q}(\alpha, i)$ is a splitting field for X^4-2 over $\mathbf{Q}$.

To determine the degree of K over $\mathbf{Q}$ we remark first that since i is a root of the second-degree polynomial X^2+1 in $P(\mathbf{Q}(\alpha))$ we have $(K: \mathbf{Q}(\alpha)) \leqq 2$. But i does not belong to $\mathbf{Q}(\alpha)$, since $\mathbf{Q}(\alpha)$ is a subfield of $\mathbf{R}$ and i is not a real number. Hence $(K: \mathbf{Q}(\alpha)) = 2$ and so $(K: \mathbf{Q}) = 8$.

§ **12. Algebraically closed fields.** Let C be a field and consider the following four properties which C may enjoy:

C1. Every non-constant polynomial in $P(C)$ has at least one root in C.

C2. Every non-constant polynomial in $P(C)$ splits completely in $P(C)$.

C3. Every non-constant irreducible polynomial in $P(C)$ is linear.

C4. If $(E, \imath)$ is an algebraic extension of C then $\imath$ is an isomorphism, i.e., $\imath(C) = E$.

It is easy to convince oneself that these four properties are equivalent; a field which enjoys one (and hence all) of them is said to be **algebraically closed.** It is proved in the theory of functions of a complex variable that the field $\mathbf{C}$ of complex numbers is algebraically closed. Although this result is traditionally known as the Fundamental Theorem of Algebra, it is essentially a theorem of analysis, and is certainly by no means fundamental to modern algebra.

Let F be a field. An extension $(C, \imath)$ of F is called an **algebraic closure** of F if it is both algebraic over F and also algebraically closed. Roughly speaking, an algebraic closure of a field is a "smallest" algebraically closed extension. For, if $(C, \imath)$ is an algebraic closure of F and K is a subfield of C, distinct from C, containing $\imath(F)$, there exists an element α of C not in K. Since α is algebraic over $\imath(F)$ it is algebraic over K and its minimum polynomial over K is a non-linear irreducible polynomial in $P(K)$; so K is not algebraically closed.

The terminology just introduced must be distinguished carefully from the " relative " terminology of § 9. One may make the distinction informally by saying that a subfield F of E is relatively algebraically closed in E if it has no algebraic extensions in E, while a field C is algebraically closed if it has no algebraic extensions at all (except for " trivial " extensions, i.e., fields isomorphic to C).

THEOREM 12.1. *An algebraic extension* (C, ι) *of a field* F *is an algebraic closure of* F *if and only if every non-constant polynomial in* $P(\iota(F))$ *splits completely in* $P(C)$.

Proof. Let us identify F and $\iota(F)$, so that we may consider F as a subfield of C and $P(F)$ as a subring of $P(C)$.

If C is an algebraic closure of F then it is algebraically closed and so every non-constant polynomial in $P(C)$—and hence every non-constant polynomial in $P(F)$—splits completely in $P(C)$.

Conversely, suppose that every non-constant polynomial in $P(F)$ splits completely in $P(C)$. We shall show that C is algebraically closed. So let f be a non-constant polynomial in $P(C)$. According to Theorem 10.5 there exists an extension (K, ι') of C such that $\iota'_P(f)$ has a root α in K. We shall identify C and $\iota'(C)$, so we may say that f has a root α in K. Let L be the subfield of C generated over F by the coefficients of f; since all these coefficients are algebraic over F, it follows from Theorem 9.2 that the degree $(L\colon F)$ is finite. Since α is a root of the polynomial f in $P(L)$, the degree $(L(\alpha)\colon L)$ is also finite. Hence $(L(\alpha)\colon F) = (L(\alpha)\colon L)(L\colon F)$ is finite, and so, by Theorem 9.1, α is algebraic over F—say α is a root of the polynomial g in $P(F)$. By hypothesis, g splits completely in $P(C)$ and hence all its roots lie in C. In particular, α belongs to C. Thus f has at least one root in C and hence C is algebraically closed.

In Theorem 10.5 we showed how to construct an extension of a field F in which a given polynomial with coefficients in F had at least one root; in Theorem 11.1 we did even better, by constructing an extension in which the given polynomial split completely. An algebraic closure of F is clearly better still, for in an algebraic closure not only a single given polynomial but every polynomial with coefficients in F splits completely. We ought now to prove existence and " essential uniqueness " theorems for algebraic closures analogous to those of §§ 10 and 11. These theorems are in fact true, but unfortunately the set-theoretic machinery involved in their proofs (transfinite induction or some equivalent procedure) is beyond the scope of this book. We content ourselves with the statements of the results.

THEOREM 12.2. *Let F be any field. Then there exists an algebraic closure (C, ι) of F. Further, if τ is an isomorphism of F onto a field F' and (C', ι') is an algebraic closure of F', there exists an isomorphism τ_1 of C onto C' such that $\tau_1(\iota(a)) = \iota'(\tau(a))$ for every element a of F.*

The second part of Theorem 12.2 is a special case of the following result which we also state without proof.

THEOREM 12.3. *Let F be a field, (C, ι) an algebraic closure of F, (E, ι') any algebraic extension of F. Then there exists a monomorphism ρ of E into C such that $\rho(\iota'(a)) = \iota(a)$ for every element a of F.*

If we identify E with its image under ρ we may summarise Theorem 12.3 by saying that every algebraic extension of F can be " embedded " in an algebraic closure of F.

§ **13. Separable extensions.** Let F be a field, f a polynomial with coefficients in F. Let (K, ι) be a splitting field for f over F. We shall identify F with $\iota(F)$ in the usual way and shall consider F as a subfield of K and $P(F)$ as a subring of $P(K)$. For each root α of f in K we define the **multiplicity** k_α of α to be the largest of the integers k such that $(X-\alpha)^k$ is a factor of f in $P(K)$. If k_α is greater than 1 we say that α is a **repeated root** of f. Clearly α is a root of f with multiplicity k_α if and only if there exists a polynomial g in $P(K)$ such that $f = (X-\alpha)^{k_\alpha}g$ and $\sigma_\alpha(g) \neq 0$.

If f has leading coefficient a and roots $\alpha_1, \ldots, \alpha_r$ with multiplicities $k_1, \ldots, k_r$ respectively then f splits into factors in $P(K)$ in the form

$$f = a(X-\alpha_1)^{k_1}(X-\alpha_2)^{k_2}\ldots(X-\alpha_r)^{k_r}.$$

From this we deduce that $\sum_{i=1}^{r} k_i = \partial f$; this result is usually expressed by saying that " when the roots are counted in their multiplicities " a polynomial of degree n has n roots in a splitting field. This implies, of course, that a polynomial of degree n cannot have more than n roots in any field.

We now give a criterion which allows us to determine whether a given polynomial has any repeated roots.

THEOREM 13.1. *Let f be a polynomial with coefficients in a field F. Let h be the highest common factor of f and its derivative Df. Then f has a repeated root in a splitting field if and only if ∂h is positive.*

Proof. (1) Suppose f has a repeated root α of multiplicity k $(k>1)$ in its splitting field K. Then there is a polynomial g such that $f = (X-\alpha)^k g$ and $\sigma_\alpha(g) \neq 0$. Computing the derivative of f according to the rules in § 5, we have

$$Df = (X-\alpha)^k Dg + k(X-\alpha)^{k-1}g.$$

So $\sigma_\alpha(f) = \sigma_\alpha(Df) = 0$. Thus f and Df, which are both polynomials in $P(F)$, are multiples in $P(F)$ of the minimum polynomial of α over F. That is to say, f and Df have a non-constant common factor in $P(F)$; hence ∂h is positive.

(2) Conversely, suppose ∂h is positive. Since h is a non-constant factor of f it splits completely in $P(K)$ and each of its linear factors in $P(K)$ is also a factor of f and Df in $P(K)$. Let $X-\alpha$ be one of those factors; we shall show that α is a repeated root of f. Suppose, to the contrary, that the multiplicity of α as root of f is 1. Then we can write $f = (X-\alpha)g$ where $\sigma_\alpha(g) \neq 0$. We then have $Df = g+(X-\alpha)Dg$, whence $\sigma_\alpha(Df) = \sigma_\alpha(g) \neq 0$, which is a contradiction, since $X-\alpha$ is a factor of Df. Thus α is a repeated root of f as asserted.

We notice that although the roots of f lie in the splitting field K, we may decide whether any of the roots is repeated by means of a procedure carried out entirely in $P(F)$.

COROLLARY. *If F has characteristic zero then no irreducible polynomial in $P(F)$ has a repeated root in a splitting field. If F has non-zero characteristic p an irreducible polynomial in $P(F)$ has repeated roots in a splitting field if and only if it has the form $a_0+a_1X^p+a_2X^{2p}+\ldots+a_nX^{np}(n \geqq 1)$, where $a_0, a_1, \ldots, a_n$ are elements of F.*

Proof. Let f be an irreducible polynomial in $P(F)$. If its derivative Df is not the zero polynomial z, the highest common factor of f and Df is the identity polynomial e; hence f has no repeated roots. On the other hand, if $Df = z$, f is itself the highest common factor of f and Df; hence, if f is non-constant, it has repeated roots in a splitting field. The result now follows at once from our characterisation in Theorem 5.3 of the polynomials in $P(F)$ which have zero derivative.

An irreducible polynomial with coefficients in a field F is said to be **separable** if it has no repeated roots in a splitting field; an arbitrary polynomial is said to be separable if all its irreducible factors are separable. (We notice that a separable polynomial may itself have repeated roots in a splitting field: but its irreducible factors do not. For example, if a is any element of F, the polynomial $(X-a)^2$ certainly has a repeated root, but it is separable since its only irreducible factor $X-a$ does not.) Polynomials which are not separable are naturally said to be **inseparable.** A field F is said to be **perfect** if there are no inseparable polynomials with coefficients in F.

Let F be a subfield of a field E. An element α of E which is algebraic over F is said to be **separable** over F if its minimum polynomial $m_{\alpha, F}$ over F is separable and **inseparable** over F if $m_{\alpha, F}$ is inseparable. An algebraic extension (E, ι) of a field F is called a **separable extension** if every element of E is separable over $\iota(F)$.

The following theorem provides us with a useful tool for our investigations of perfect fields and separable extensions.

THEOREM 13.2. *Let F be a field of non-zero characteristic p, E an algebraic extension of F containing F, α an element of E. If α is separable over $F(\alpha^p)$, then α actually belongs to $F(\alpha^p)$.*

Proof. Let us write $F_1 = F(\alpha^p)$.

Let m be the minimum polynomial of α over F_1; by hypothesis, m is separable. Since α is a root of the polynomial $X^p - \alpha^p$ in $P(F_1)$, it follows from Theorem 8.2 that m is a factor of $X^p - \alpha^p$ in $P(F_1)$. In $P(E)$ we have the decomposition $X^p - \alpha^p = (X-\alpha)^p$. So, according to Theorem 10.4, we may deduce that m is a power of $X-\alpha$. But since m is a separable irreducible polynomial it can have no repeated roots; hence $m = X-\alpha$. Since m is in $P(F_1)$ it follows that α belongs to F_1 as required.

Our first application of Theorem 13.2 is to give a criterion for a field to be perfect.

THEOREM 13.3. *All fields of characteristic zero are perfect. A field F of non-zero characteristic p is perfect* *if and only if every element of F is the* pth *power of some element of F.*

Proof. The first statement follows at once from the Corollary to Theorem 13.1.

Let now F be a field of non-zero characteristic p.

(1) Suppose F is perfect. Let a be any element of F; we claim that there is an element b in F such that $a = b^p$.

Let f be the polynomial $X^p - a$. According to Theorem 10.5 there exists an extension (E, ι) of F such that $\iota_P(f)$ has a root α in E. We shall identify F and $\iota(F)$ in the usual way; then $\alpha^p = a$ and so $F(\alpha^p) = F(a) = F$. Since F is perfect, α is separable over F and hence, by Theorem 13.2, α actually belongs to F, and since $\alpha^p = a$, α is the required element b.

(2) Conversely, suppose that every element of F is the pth power of an element of F.

Suppose there is an irreducible polynomial f in $P(F)$ which has repeated roots in a splitting field. Then, according to the Corollary to Theorem 13.1, f has the form $a_0 + a_1 X^p + a_2 X^{2p} + \ldots + a_n X^{np}$, where $a_0, \ldots, a_n$ are elements of F. By hypothesis, there exist elements $b_0, \ldots, b_n$ of F such that $a_i = b_i^p (i = 1, \ldots, n)$. Hence, since F has characteristic p, $f = (b_0 + b_1 X + b_2 X^2 + \ldots + b_n X^n)^p$ and so is not irreducible. This is a contradiction; so it follows that no irreducible polynomial in $P(F)$ can have repeated roots in a splitting field. Thus all irreducible polynomials —and hence all polynomials—in $P(F)$ are separable; that is to say, F is perfect.

COROLLARY. *Every finite field is perfect.*

Proof. Let F be a finite field of characteristic p; p is of course non-zero (Corollary to Theorem 3.2). We saw in Theorem 3.3 that the mapping π of F into itself defined by setting $\pi(x) = x^p$ for all elements x of F is one-to-one. Hence the number of images under π in F is equal to the number of elements of F. That is to say, every element of F is an image under π, i.e., is the pth power of some element of F. The result now follows from Theorem 14.3.

Our next statement is an immediate consequence of the definition of a separable extension.

THEOREM 13.4. *Let D be an algebraic extension of a subfield F, E a subfield of D including F. If D is a separable extension of F, then E is a separable extension of F and D is a separable extension of E.*

Proof. The first statement is obvious.

Let now α be any element of D, m_F and m_E the minimum polynomials of α over F and E respectively. Then m_E is a factor of m_F in $P(E)$. Since α is separable over F, m_F is a separable polynomial; hence so also is m_E. Thus α is separable over E. It follows that D is a separable extension of E as asserted.

We now propose to establish the converse of Theorem 13.4, at least for extensions of finite degree. To this end we shall first obtain a necessary and sufficient condition for an algebraic extension to be separable. Since all algebraic extensions of fields of characteristic zero are separable (Theorem 13.3), we can concentrate our attention on fields of non-zero characteristic.

So let E be an extension of finite degree over a subfield F of non-zero characteristic p. Let π be the monomorphism of E into itself defined by setting $\pi(x) = x^p$ for every element x of E. We shall denote the image $\pi(E)$ of E under π by E^p; this is a sensible notation, since $\pi(E)$ is the subfield of E consisting of the pth powers of elements of E. Let $L = F(E^p)$ be the subfield of E obtained by adjoining E^p to F. Since E is a finite extension of F, so also is L and hence L can be generated over F by a finite subset of E^p, say $L = F(\alpha_1^p, \ldots, \alpha_n^p)$, where $\alpha_1, \ldots, \alpha_n$ are suitably chosen elements of E.

It follows from Theorem 9.1 that α_1^p is algebraic over F, and so, according to Theorem 8.2, we can deduce that the elements of $F(\alpha_1^p)$ can be expressed as polynomials in α_1^p with coefficients in F. Similarly, since α_2^p is algebraic over F and hence over $F(\alpha_1^p)$, every element of $F(\alpha_1^p, \alpha_2^p)$ can be expressed as a polynomial in α_2^p with coefficients in $F(\alpha_1^p)$, and so has the form

$$\sum_j \left(\sum_i a_{ij}\alpha_1^{pi}\right)\alpha_2^{pj} = \sum_i \sum_j a_{ij}(\alpha_1^i \alpha_2^j)^p$$

with a_{ij} in F. Proceeding in this way we see that every element of L can be expressed as a finite linear combination of elements of E^p with coefficients in F. Of course all such linear combinations belong to L. Thus L consists precisely of those elements of E which can be expressed as finite linear combinations of elements of E^p with coefficients in F.

We are now in a position to establish a new criterion for separability.

THEOREM 13.5. *Let E be an extension of finite degree over a subfield F of non-zero characteristic p. Then E is a separable extension of F if and only if $F(E^p) = E$.*

Proof. As above, we shall write $L = F(E^p)$.

(1) Suppose E is a separable extension of F.

Let α be any element of E. Then α is separable over F and hence, by Theorem 13.4, over $F(\alpha^p)$. According to Theorem 13.2, this implies that α belongs to $F(\alpha^p)$, which is of course a subfield of L.

Thus α belongs to L; hence $E = L$, as asserted.

(2) Conversely, suppose that $E = L$. Set $(E\colon F) = n$.

We show first that if $\{x_1, \ldots, x_k\}$ is a subset of E linearly independent over F then so is $\{x_1^p, \ldots, x_k^p\}$. To this end we complete $\{x_1, \ldots, x_k\}$ to a basis $\{x_1, \ldots, x_n\}$ of E over F (Theorem 4.4) and remark that then every element x of E

can be expressed in the form $x = a_1x_1 + \ldots + a_nx_n$, where $a_1, \ldots, a_n$ are elements of F; then

$$\pi(x) = x^p = a_1^p x_1^p + \ldots + a_n^p x_n^p.$$

It now follows from the remarks immediately before this theorem that every element of L has the form

$$b_1 x_1^p + \ldots + b_n x_n^p,$$

where $b_1, \ldots, b_n$ are elements of F. Thus the set $\{x_1^p, \ldots, x_n^p\}$ is a generating system over F for the n-dimensional vector space $E(= L)$, and hence (by Theorem 4.3) is linearly independent over F; hence its subset $\{x_1^p, \ldots, x_k^p\}$ is also linearly independent as asserted.

Let now α be any element of E. Since $(E: F)$ is finite, it follows from Theorem 9.1 that α is algebraic over F; let m be the minimum polynomial of α over F. Suppose that m is inseparable; then, since m is irreducible, the Corollary to Theorem 13.1 shows that m has the form

$$m = c_0 + c_1 X^p + c_2 X^{2p} + \ldots + c_r X^{rp},$$

where $c_0, \ldots, c_r$ are elements of F, not all zero, and $rp \leqq n$, since $rp = \partial m = (F(\alpha): F) \leqq (E: F) = n$. Thus we have

$$c_0 + c_1\alpha^p + c_2\alpha^{2p} + \ldots + c_r\alpha^{rp} = 0,$$

so that the set $\{e, \alpha^p, \alpha^{2p}, \ldots, \alpha^{rp}\}$ is linearly dependent over F. But since the set $\{e, \alpha, \ldots, \alpha^{rp-1}\}$ is linearly independent over F (Theorem 8.2), its subset $\{e, \alpha, \ldots, \alpha^r\}$ is linearly independent and hence, by the preceding paragraph, $\{e, \alpha^p, \ldots, \alpha^{rp}\}$ is linearly independent over F. The hypothesis that m is inseparable has thus led to a contradiction.

It follows that E is a separable extension of F, as asserted.

We can now establish the converse of Theorem 13.4.

THEOREM 13.6. *Let D be an extension of finite degree over a subfield F, E a subfield of D including F. If E is a separable extension of F and D is a separable extension of E, then D is a separable extension of F.*

Proof. The result is clear when F has characteristic zero, since in this case all extensions are separable.

So suppose that F has non-zero characteristic p. Then it follows from the first part of Theorem 13.5 that $E = F(E^p)$ and $D = E(D^p)$. Hence D can be obtained by adjoining E^p and D^p to F; but since E is a subfield of D, E^p is a subfield of D^p, and hence D is obtained simply by adjoining D^p to F, i.e., $D = F(D^p)$. Hence, according to the second part of Theorem 13.5, D is a separable extension of F.

We show next that if a field E is obtained by adjoining a finite set of elements $\{x_1, ..., x_n\}$ to a subfield F then, in order to determine whether E is a separable extension of F, we do not have to examine every element x of E for separability over F: it is enough to examine the generating elements $x_1, ..., x_n$.

THEOREM 13.7. *Let E be a field generated over a subfield F by a finite set of elements $x_1, ..., x_n$ all of which are separable over F. Then E is a separable extension of F.*

Proof. Again the result is clear when F has characteristic zero; so we suppose that F has non-zero characteristic p.

We proceed by induction on n. The result is trivially true when $n = 0$. So suppose we have established that if $x_1, ..., x_k$ are separable over F, then $E_k = F(x_1, ..., x_k)$ is separable over F.

Let now x_{k+1} be an element of E which is separable over F; set $E_{k+1} = E_k(x_{k+1}) = F(x_1, ..., x_{k+1})$. We shall show that $E_{k+1} = E_k(E_{k+1}^p)$; to this end it is clearly sufficient to show that x_{k+1} belongs to $E_k(E_{k+1}^p)$. By hypothesis x_{k+1} is separable over F and hence over

$E_k(x_{k+1}^p)$; it then follows from Theorem 13.2 that x_{k+1} actually belongs to $E_k(x_{k+1}^p)$ and so to $E_k(E_{k+1}^p)$. Consequently E_{k+1} is a separable extension of E_k (Theorem 13.5). According to the inductive hypothesis, E_k is a separable extension of F; hence, by Theorem 13.6, E_{k+1} is a separable extension of F.

Thus the induction is complete and the theorem established.

COROLLARY. *Let f be a separable polynomial with coefficients in a field F. Then every splitting field of f over F which includes F is a separable extension of F.*

Proof. This follows at once from the theorem, since each splitting field is obtained by adjoining to F all the roots of f, and these roots are all separable over F.

In the remaining chapters of this book we shall be dealing exclusively with separable extensions. But, in order to show that inseparability really can occur, we conclude this section with an example of an inseparable polynomial.

It follows from Theorem 13.3 and its Corollary that we cannot hope to find an inseparable polynomial with coefficients in a field of characteristic zero nor in a finite field. So we begin with the prime field $\mathbf{Z}_2$ which has two elements and characteristic 2, and form $F = R(\mathbf{Z}_2)$, the field of rational functions with coefficients in $\mathbf{Z}_2$—this is an infinite field of characteristic 2. We denote the polynomial $(0, e, 0, 0, \ldots)$ in $P(\mathbf{Z}_2)$ by X as usual, and regard X as an element of F. Now we consider the polynomial ring $P(F)$ and denote the polynomial $(0, e_F, 0, 0, \ldots)$ in $P(F)$ by Y.

Let f be the polynomial $Y^2 - X$ in $P(F)$. We claim that f is irreducible in $P(F)$. For, if it were reducible in $P(F)$, there would be an element,

$$\alpha = \frac{a_0 + a_1 X + \ldots + a_m X^m}{b_0 + b_1 X + \ldots + b_n X^n}$$

say, in F such that $\alpha^2 = X$; then we would have

$$a_0^2+a_1^2X^2+\ldots+a_m^2X^{2m} = X(b_0^2+b_1^2X^2+\ldots+b_n^2X^{2n}).$$

Since X is transcendental over $\mathbf{Z}_2$ in F, all the coefficients must be zero, which is impossible. Hence f is irreducible in $P(F)$.

But Df is the zero polynomial; so f is inseparable.

Examples II

1. Let F be a field, and identify F and the polynomial ring $P(F)$ with subrings of the rational function field $R(F)$ as described at the end of § 6. Show that when this identification is carried out we have $R(F) = F(X)$, the simple extension generated over F by the polynomial X.

2. Show that the only monomorphism of a prime field into itself is the identity automorphism. Deduce that if a field F is of finite degree over its prime subfield then every monomorphism of F into itself is an automorphism.

3. Let ι_n be the mapping of $R(F) = F(X)$ into itself defined by setting

$$\iota_n\left(\frac{a_0+a_1X+\ldots+a_rX^r}{b_0+b_1X+\ldots+b_sX^s}\right) = \frac{a_0+a_1X^n+\ldots+a_rX^{rn}}{b_0+b_1X^n+\ldots+b_sX^{sn}}.$$

Show that ι_n is a monomorphism of $F(X)$ onto the subfield $F(X^n)$ and that $(F(X): F(X^n)) = n$. (Contrast Example 2.)

4. If F is a subfield of E and E is a subfield of K such that E is algebraic over F and K is algebraic over E, prove that K is algebraic over F. Deduce that the relative algebraic closure of a subfield F in a field L is relatively algebraically closed in L.

5. Show that a field F, considered as a subfield of its field of rational functions $R(F)$, is relatively algebraically closed in $R(F)$.

6. Let F be a field, f a polynomial in $P(F)$, not necessarily irreducible. As in the first two paragraphs of the proof of Theorem 10.1 we may define congruence of polynomials modulo f and construct the ring E of residue classes modulo f. Show that a residue class modulo f has a multiplicative inverse in E if and only if it consists of polynomials which are relatively prime to f.

7. Let F and F' be fields, τ an isomorphism of F onto F', τ_P the canonical extension of τ to $P(F)$. Let f be a non-constant polynomial in $P(F)$, not necessarily irreducible, and let (E, ι) and (E', ι') be extensions of F and F' in which f and $\tau_P(f)$ have roots β and β' respectively. Identify F and F' with their images under ι and ι' respectively. Show that there exists an extension τ_1 of τ to $F(\beta)$ such that $\tau_1(\beta) = \beta'$ if and only if $m_{\beta', F'} = \tau_P(m_{\beta, F})$.

8. Construct subfields of $\mathbf{C}$ which are splitting fields over $\mathbf{Q}$ for the polynomials X^4-5X^2+6, X^6-1, X^6-8, and find their degrees over $\mathbf{Q}$.

9. Let F be a finite field with elements $a_0 = 0, a_1, \ldots, a_k$. By considering the polynomial

$$a_1+(X-a_0)(X-a_1)\ldots(X-a_k)$$

in $P(F)$ show that F is not algebraically closed.

10. If F is a subfield of an algebraically closed field C, prove that the relative algebraic closure of F in C is in fact an algebraic closure of F. Deduce that $\mathbf{A}$ and $\mathbf{C}$ are algebraic closures of $\mathbf{Q}$ and $\mathbf{R}$ respectively.

11. Let f_1 and f_2 be polynomials with coefficients in a field F, α an element of F. If α is a root of f_1 and of f_2 with multiplicities k_1 and k_2 respectively, show that α is a root of $f_1 f_2$ with multiplicity k_1+k_2. Let k be the multiplicity of α as root of f_1+f_2; show that $k \geqq \min(k_1, k_2)$ and that if $k_1 \neq k_2$ then $k = \min(k_1, k_2)$.

12. Let f be a monic irreducible polynomial in $P(F)$, where F is a field of non-zero characteristic p. Show that there exists an integer $e \geqq 0$ such that f can be expressed in the form $\Sigma a_r X^{rp^e}$ but not in the form $\Sigma b_s X^{sp^{e+1}}$; show that the polynomial $f_0 = \Sigma a_r X^r$ is irreducible in $P(F)$ and that it has no repeated roots in a splitting field. Let K_0 be a splitting field for f_0 over F including F, K a splitting field for f over K_0 including K_0. If $\beta_1, \ldots, \beta_{n_0}$ are the roots of f_0 in K_0 show that the roots of f in K are $\alpha_1, \ldots, \alpha_{n_0}$ where $\alpha_i^{p^e} = \beta_i$ $(i = 1, \ldots, n_0)$ and that each root of f has multiplicity p^e.

CHAPTER III

GALOIS THEORY

§ **14. Automorphisms of fields.** Let K be a field, S any set, and let $M(S, K)$ be the set of all mappings of S into K. We define an operation of addition in $M(S, K)$: for every pair of mappings ϕ_1, ϕ_2 in $M(S, K)$ we define the mapping $\phi_1+\phi_2$ of S into K by setting

$$(\phi_1+\phi_2)(s) = \phi_1(s)+\phi_2(s)$$

for every element s of S. We also define an operation of multiplication of elements of $M(S, K)$ by elements of K: for every element a of K and every mapping ϕ in $M(S, K)$ we define the mapping $a\phi$ of S into K by setting

$$(a\phi)(s) = a\phi(s)$$

for every element s of S. It is easy to verify that, under the two operations we have defined, $M(S, K)$ is a vector space over the field K. In this vector space it of course makes sense to talk of linear dependence and independence. Namely, the set $\{\phi_1, \ldots, \phi_r\}$ of mappings of S into K is linearly dependent over K if there exist elements $a_1, \ldots, a_r$ of K, not all zero, such that $a_1\phi_1+\ldots+a_r\phi_r$ is the zero mapping ζ of S into K, i.e., such that

$$a_1\phi_1(s)+\ldots+a_r\phi_r(s) = \zeta(s) = 0$$

for every element s of S. Otherwise $\{\phi_1, \ldots, \phi_r\}$ is linearly independent. With this introduction we are now in a position to state and prove an important result due to Dedekind.

THEOREM 14.1. *Let E and K be fields. Every set of distinct monomorphisms of E into K is linearly independent over K.*

Proof. Since an infinite set is linearly independent if and only if every finite subset is linearly independent, we need only consider finite sets of distinct monomorphisms. We proceed by induction on the number n of monomorphisms in the set.

First suppose $n = 1$. A set consisting of a single monomorphism τ of E into K is clearly linearly independent over K. For if $a\tau(x) = 0$ for every element x of E we must have $a = 0$ (since $\tau(x_1)$ is non-zero for every non-zero element x_1 of E).

Suppose next that we have established that every set consisting of fewer than k monomorphisms of E into K is linearly independent. Let $\{\tau_1, \ldots, \tau_k\}$ be a set of k distinct monomorphisms of E into K and suppose that there exist elements $a_1, \ldots, a_k$ of K, not all zero, such that

$$a_1\tau_1 + \ldots + a_k\tau_k = \zeta. \tag{14.1}$$

Then none of the elements $a_1, \ldots, a_k$ is zero: if, for example, $a_k = 0$, (14.1) would imply that the set $\{\tau_1, \ldots, \tau_{k-1}\}$ is linearly dependent, in contradiction to our inductive hypothesis. Dividing (14.1) by the non-zero element a_k we obtain a relation of the form

$$b_1\tau_1 + \ldots + b_{k-1}\tau_{k-1} + \tau_k = \zeta \tag{14.2}$$

with non-zero coefficients $b_1, \ldots, b_{k-1}$ in K.

Since τ_1 and τ_k are distinct monomorphisms there exists an element x_1 of E such that $\tau_1(x_1) \neq \tau_k(x_1)$; x_1 is clearly non-zero. Then for every element x of E we have

$$b_1\tau_1(x_1x) + \ldots + b_{k-1}\tau_{k-1}(x_1x) + \tau_k(x_1x) = 0,$$

and hence, since $\tau_1, \ldots, \tau_k$ are monomorphisms, it follows that, for every element x of E,

$$b_1\tau_1(x_1)\tau_1(x)+\ldots+b_{k-1}\tau_{k-1}(x_1)\tau_{k-1}(x)+\tau_k(x_1)\tau_k(x) = 0. \tag{14.3}$$

We may divide (14.3) by the non-zero element $\tau_k(x_1)$ and so obtain the relation

$$b_1\frac{\tau_1(x_1)}{\tau_k(x_1)}\tau_1+\ldots+b_{k-1}\frac{\tau_{k-1}(x_1)}{\tau_k(x_1)}\tau_{k-1}+\tau_k=\zeta. \tag{14.4}$$

Subtracting (14.4) from (14.2) we obtain

$$b_1\left(e-\frac{\tau_1(x_1)}{\tau_k(x_1)}\right)\tau_1+\ldots+b_{k-1}\left(e-\frac{\tau_{k-1}(x_1)}{\tau_k(x_1)}\right)\tau_{k-1}=\zeta, \tag{14.5}$$

where e is the identity of K. Now x_1 was chosen in such a way that the coefficient of τ_1 in (14.5) is non-zero. Hence (14.5) shows that the set $\{\tau_1, \ldots, \tau_{k-1}\}$ is linearly dependent. But this contradicts our inductive hypothesis.

So the induction is complete and the theorem proved.

Let K be a field and let A be the set of automorphisms of K; we proceed to define a law of composition on A and shall show that A, equipped with this law of composition, is a group. Namely, let τ_1 and τ_2 be automorphisms of K; then we define the mapping $\tau_1\tau_2$ of K into itself by setting

$$(\tau_1\tau_2)(x) = \tau_1(\tau_2(x))$$

for all elements x of K. We claim that $\tau_1\tau_2$ is an automorphism of K. First, let x and y be any two elements of K; then since τ_1 and τ_2 are homomorphisms, we have

$$\begin{aligned}(\tau_1\tau_2)(xy) &= \tau_1(\tau_2(xy)) = \tau_1(\tau_2(x)\tau_2(y)) \\ &= \tau_1(\tau_2(x))\tau_1(\tau_2(y)) = (\tau_1\tau_2)(x)(\tau_1\tau_2)(y)\end{aligned}$$

and similarly

$$(\tau_1\tau_2)(x+y) = (\tau_1\tau_2)(x)+(\tau_1\tau_2)(y);$$

so $\tau_1\tau_2$ is a homomorphism of K into itself. Next we show that $\tau_1\tau_2$ is one-to-one. Suppose that

$$(\tau_1\tau_2)(x) = (\tau_1\tau_2)(y);$$

thus $\tau_1(\tau_2(x)) = \tau_1(\tau_2(y))$, whence $\tau_2(x) = \tau_2(y)$, and hence $x = y$, since τ_1 and τ_2 are both one-to-one. Finally we show that $\tau_1\tau_2$ maps K *onto* K. Let z be any element of K; since τ_1 is an epimorphism, there is an element y of K such that $\tau_1(y) = z$, and, since τ_2 is an epimorphism, there is an element x of K such that $\tau_2(x) = y$; thus $z = (\tau_1\tau_2)(x)$. It follows that the mapping $\tau_1\tau_2$ is an automorphism of K; so we have actually defined a law of composition on A.

We now show that A, equipped with this law of composition, is a group. To this end we must first establish that the law of composition is associative; so let τ_1, τ_2, τ_3 be elements of A, and x any element of K. We have

$$((\tau_1\tau_2)\tau_3)(x) = (\tau_1\tau_2)(\tau_3(x)) = \tau_1(\tau_2(\tau_3(x))),$$

$$(\tau_1(\tau_2\tau_3))(x) = \tau_1((\tau_2\tau_3)(x)) = \tau_1(\tau_2(\tau_3(x)));$$

thus $(\tau_1\tau_2)\tau_3 = \tau_1(\tau_2\tau_3)$, i.e., the law of composition is associative. The identity automorphism ε of K, defined by setting $\varepsilon(x) = x$ for all elements x of K, is clearly a neutral element for the law of composition. Lastly we must show that every element of A has an inverse relative to the law of composition. Let τ be any element of A; since τ is a one-to-one mapping of A onto itself, we can define a mapping τ^{-1} of K into itself by setting, for every element x of K, $\tau^{-1}(x) = x'$, where x' is the unique element of K whose image under τ is x. (There is at least one such element since τ maps K onto K, and exactly one since τ is one-to-one.) We claim that the mapping τ^{-1} is an automorphism of K. It is clearly a one-to-one mapping of K onto itself; and it is a homomorphism, for if x and y are any two elements of K and $\tau^{-1}(x) = x'$, $\tau^{-1}(y) = y'$, then $\tau(x'y') = \tau(x')\tau(y') = xy$ and so we have

$$\tau^{-1}(xy) = x'y' = \tau^{-1}(x)\tau^{-1}(y).$$

It follows at once from the definition of τ^{-1} that $\tau\tau^{-1} = \tau^{-1}\tau = \varepsilon$. Thus A, with the law of composition defined above, is a group; we call it the **automorphism group** of K.

Let K_1 and K_2 be fields with a common subfield F. Let τ be a monomorphism of K_1 into K_2, x an element of F. We say that τ **acts like the identity on** x or that τ **leaves** x **fixed** if $\tau(x) = x$; we say that τ acts like the identity on F, or leaves F fixed, if it leaves every element of F fixed. If τ leaves F fixed we often call it an ***F*-monomorphism**; if τ is an isomorphism of K_1 onto K_2 which leaves F fixed, we call it an ***F*-isomorphism**; in particular, if $K_1 = K_2$ and τ is an automorphism of K_1 which leaves F fixed, we call τ an ***F*-automorphism.** There is a point about this terminology which we must make quite clear: when we say that a mapping τ is an F-monomorphism we are not asserting that the set of elements left fixed by τ is precisely F, but merely that F is included in this set.

Let G be a subgroup of the automorphism group of a field K; let F_0 be the subset of K consisting of those elements of K which are left fixed by every automorphism in G. It is easily verified that F_0 is a subfield of K; we call it the **fixed field under** G.

THEOREM 14.2. *Let G be a finite subgroup of the automorphism group of a field K, F_0 the fixed field under G. Then the degree of K over F_0 is equal to the order of the group G.*

Proof. Suppose G has order n; let $\tau_1 = \varepsilon, \tau_2, \ldots, \tau_n$ be the elements of G. The proof proceeds in two stages: we prove first that $(K: F_0) \geqq n$ and then that $(K: F_0) \leqq n$.

First suppose that $(K: F_0) = m < n$. Let $\{x_1, \ldots, x_m\}$ be a basis for K over F_0. Since a system of homogeneous linear equations with more unknowns than equations always has a non-trivial solution it follows that there exist elements $y_1, \ldots, y_n$ of K, not all zero, such that

$$\tau_1(x_j)y_1+\ldots+\tau_n(x_j)y_n = 0 \quad (j = 1, \ldots, m). \qquad (14.6)$$

Let a be any element of K; then there exist elements $\alpha_1, \ldots, \alpha_m$ of F_0 such that $a = \alpha_1 x_1+\ldots+\alpha_m x_m$. Multiply the equations (14.6) by $\alpha_1, \ldots, \alpha_m$; remembering that $\alpha_j = \tau_1(\alpha_j) = \tau_2(\alpha_j) = \ldots = \tau_n(\alpha_j)$, we obtain

$$\tau_1(\alpha_j)\tau_1(x_j)y_1+\ldots+\tau_n(\alpha_j)\tau_n(x_j)y_n = 0 \quad (j = 1, \ldots, m).$$

Since $\tau_1, \ldots, \tau_n$ are automorphisms of K these equations may be written

$$\tau_1(\alpha_j x_j)y_1+\ldots+\tau_n(\alpha_j x_j)y_n = 0 \quad (j = 1, \ldots, m).$$

If we add these equations and recall that

$$\tau_i(\alpha_1 x_1)+\ldots+\tau_i(\alpha_m x_m) = \tau_i(\alpha_1 x_1+\ldots+\alpha_m x_m) = \tau_i(a) \qquad (i = 1, \ldots, n),$$

we obtain the result that

$$y_1\tau_1(a)+\ldots+y_n\tau_n(a) = 0$$

for every element a of K. Thus the set $\{\tau_1, \ldots, \tau_n\}$ of distinct automorphisms of K is linearly dependent over K, contrary to Theorem 14.1.

So we have established that $(K: F_0) \geqq n$.

Now suppose that $(K: F_0) > n$. Then there exists a set consisting of $n+1$ elements of K, $\{x_1, \ldots, x_{n+1}\}$ say, which is linearly independent over F_0. Further, since a system of n homogeneous linear equations with $n+1$ unknowns always has a non-trivial solution, there exist elements $y_1, \ldots, y_{n+1}$ of K, not all zero, such that

$$\tau_j(x_1)y_1+\ldots+\tau_j(x_{n+1})y_{n+1} = 0 \quad (j = 1, \ldots, n). \qquad (14.7)$$

Let us suppose that the elements $y_1, \ldots, y_{n+1}$ are so chosen that as few as possible are non-zero; that is to say, if

$z_1, \ldots, z_{n+1}$ are elements of K, not all zero, such that

$$\tau_j(x_1)z_1 + \ldots + \tau_j(x_{n+1})z_{n+1} = 0 \quad (j = 1, \ldots, n), \quad (14.8)$$

then there are at least as many non-zero elements among $\{z_1, \ldots, z_{n+1}\}$ as there are among $\{y_1, \ldots, y_{n+1}\}$. By suitably renumbering we can arrange matters in such a way that $y_1, \ldots, y_r$ are non-zero and $y_{r+1}, \ldots, y_{n+1}$ are zero.

The equations (14.7) now become

$$\tau_j(x_1)y_1 + \ldots + \tau_j(x_r)y_r = 0 \quad (j = 1, \ldots, n).$$

Dividing by y_r and setting $y_i' = y_i/y_r$ $(i = 1, \ldots, r-1)$ we obtain

$$\tau_j(x_1)y_1' + \ldots + \tau_j(x_{r-1})y_{r-1}' + \tau_j(x_r) = 0 \quad (j = 1, \ldots, n). \quad (14.9)$$

Since $\tau_1 = \varepsilon$, the identity automorphism, the first of these equations is simply

$$x_1 y_1' + \ldots + x_{r-1}y_{r-1}' + x_r = 0. \quad (14.10)$$

It follows that the elements $y_1', \ldots, y_{r-1}'$ do not all belong to F_0; if they did, (14.10) would imply that $\{x_1, \ldots, x_r\}$ is linearly dependent over F_0. Suppose y_1' does not belong to F_0. (We have shown, incidentally, that $r \neq 1$; for if $r = 1$, (14.10) reduces to $x_1 = 0$, which is impossible, since x_1 belongs to a linearly independent set.)

Since y_1' does not belong to F_0 it is not left fixed by all the automorphisms in G: thus there exists an automorphism—say τ_2—in G such that $\tau_2(y_1') \neq y_1'$. Apply τ_2 to all the equations (14.9): we obtain

$$(\tau_2\tau_j)(x_1)\tau_2(y_1') + \ldots + (\tau_2\tau_j)(x_{r-1})\tau_2(y_{r-1}') + (\tau_2\tau_j)(x_r) = 0 \quad (j = 1, \ldots, n). \quad (14.11)$$

Since G is a group, the set $\{\tau_2\tau_1, \tau_2\tau_2, \ldots, \tau_2\tau_n\}$ coincides with the set $\{\tau_1, \tau_2, \ldots, \tau_n\}$, though the order of the elements will be different. So equations (14.11) may be rewritten

$$\tau_j(x_1)\tau_2(y_1')+\ldots+\tau_j(x_{r-1})\tau_2(y_{r-1}')+\tau_j(x_r)=0 \qquad (j=1, \ldots, n). \tag{14.12}$$

Subtracting (14.12) from (14.9) we have

$$\tau_j(x_1)(y_1'-\tau_2(y_1'))+\ldots+\tau_j(x_{r-1})(y_{r-1}'-\tau_2(y_{r-1}'))=0 \qquad (j=1, \ldots, n).$$

If we set $z_i = y_i'-\tau_2(y_i')$ $(i = 1, \ldots, r-1)$, and $z_i = 0$ $(i = r, \ldots, n+1)$, then $z_1 \neq 0$ and so $z_1, \ldots, z_{n+1}$ are elements of K, not all zero, satisfying (14.8), but with fewer than r non-zero elements among them. This, however, contradicts our choice of the set $\{y_1, \ldots, y_{n+1}\}$.

Hence $(K: F_0) = n$, as asserted.

§ **15. Normal extensions.** Let (K, ι) be an algebraic extension of a field F. We shall identify F with its image under ι; so we consider F as a subfield of K and regard polynomials with coefficients in F as polynomials with coefficients in K. We say that K is a **normal extension** of F if every irreducible polynomial in $P(F)$ which has one root in K splits completely in $P(K)$. The following theorem shows that normal extensions of finite degree are easily constructed.

THEOREM 15.1. *Let K be an algebraic extension of finite degree over a subfield F. Then K is a normal extension of F if and only if K is a splitting field over F of some polynomial in $P(F)$.*

Proof. (1) Suppose K is a normal extension of F.

Let $\{x_1, \ldots, x_r\}$ be a basis for K over F, and let $m_1, \ldots, m_r$ be the minimum polynomials of $x_1, \ldots, x_r$ respectively over F. Each of the polynomials m_i has a root—namely x_i—in K, and hence, since K is a normal extension of F, each of these polynomials splits completely in $P(K)$. Let $f = m_1 m_2 \ldots m_r$; then f splits completely in $P(K)$. Now let E be any subfield of K including F such that f splits completely in $P(E)$. Then $x_1, \ldots, x_r$ belong to E, and hence E coincides with K. Thus K is a splitting field over F of the polynomial f.

(2) Conversely, suppose K is a splitting field over F of a polynomial f in $P(F)$, which we may, without loss of generality, suppose to be monic. Then $K = F(\alpha_1, \ldots, \alpha_k)$ where $\alpha_1, \ldots, \alpha_k$ are the roots of f in K.

Let p be an irreducible polynomial in $P(F)$ which has a root β in K. We may consider p as a polynomial in $P(K)$—no longer irreducible, of course—and construct a splitting field L for p over K. (We shall suppose the usual identification made, so that K may be regarded as a subfield of L.) Let β' be another root of p in L; we shall show that β' actually lies in K.

Since p is irreducible in $P(F)$ it follows from Theorem 10.2 (taking τ to be the identity automorphism of F) that there exists an isomorphism τ_1 of $F(\beta)$ onto $F(\beta')$ such that $\tau_1(\beta) = \beta'$ and $\tau_1(a) = a$ for all elements a of F. Now the subfields $K(\beta) = K$ and $K(\beta')$ of L are clearly splitting fields over $F(\beta)$ and $F(\beta')$ respectively of the original polynomial f. Hence, according to Theorem 11.2, there is an isomorphism τ' of K onto $K(\beta')$ which acts like τ_1 on $F(\beta)$, i.e., such that $\tau'(\beta) = \beta'$ and $\tau'(a) = a$ for all elements a of F. The relations between the various fields are illustrated by the commutative diagram in fig. 4; unlettered arrows represent the appropriate inclusion monomorphisms.

Since the coefficients of f lie in F, it follows that $\tau'_P(f) = f$. Now in $P(K)$ we have

$$f = (X-\alpha_1)(X-\alpha_2)\ldots(X-\alpha_k);$$

hence, applying τ'_P, we obtain

$$f = \tau'_P(f) = (X - \tau'(\alpha_1))(X - \tau'(\alpha_2))\dots(X - \tau'(\alpha_k)).$$

Thus we have two decompositions of f into linear factors in $P(L)$. As we saw in § 10, any two such decompositions differ only in the order of the factors. Hence the sets

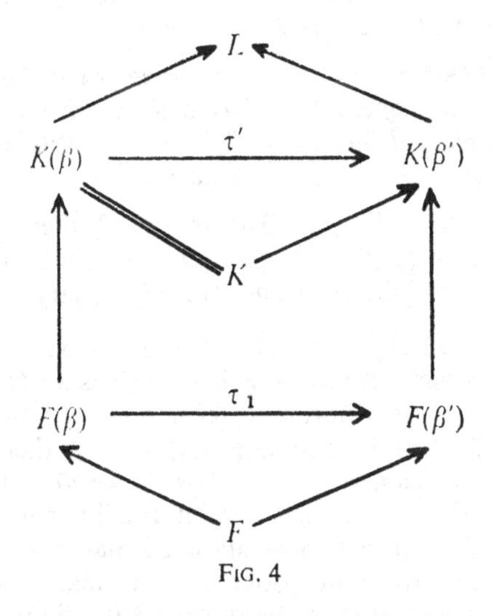

Fig. 4

$\{\alpha_1, \dots, \alpha_k\}$ and $\{\tau'(\alpha_1), \dots, \tau'(\alpha_k)\}$ consist of precisely the same elements. We now conclude that

$$K(\beta') = \tau'(K) = \tau'(F(\alpha_1, \dots, \alpha_k)) = F(\tau'(\alpha_1), \dots, \tau'(\alpha_k))$$
$$= F(\alpha_1, \dots, \alpha_k) = K,$$

and hence that β' lies in K.

So K is a normal extension of F.

Corollary 1. *Let K be a normal extension of finite degree over a subfield F. If E is any subfield of K including F then K is a normal extension of E.*

COROLLARY 2. *Let K be a normal extension of finite degree over a subfield F. If E is a subfield of K including F, and τ_1 is an F-monomorphism of E into K, then there exists an F-automorphism of K which acts like τ_1 on E.*

Proof. Since K is a normal extension of F it is a splitting field over F of some polynomial p in $P(F)$. Then K is also a splitting field for p over E and over $\tau_1(E)$. The result now follows from Theorem 11.2.

COROLLARY 3. *Let K be a normal extension of finite degree over a subfield F. If α_1 and α_2 are roots in K of an irreducible polynomial in $P(F)$ there exists an F-automorphism of K which maps α_1 onto α_2.*

Proof. According to Theorem 10.2 there exists an F-isomorphism τ_1 of $F(\alpha_1)$ onto $F(\alpha_2)$ such that $\tau_1(\alpha_1) = \alpha_2$. The result then follows at once from Corollary 2.

Let E be an algebraic extension of a subfield F. Then a **normal closure** of E over F is an extension (N, ι) of E such that (1) N is a normal extension of $\iota(F)$ and (2) if L is any subfield of N including $\iota(E)$, $L \neq N$, then L is not a normal extension of $\iota(F)$. Thus, if we identify E with its image under ι we may say informally that a normal closure of E over F is a smallest normal extension of F including E. We now prove that normal closures can actually be constructed, at least when E is a finite extension of F, and that the construction leads to an essentially unique result.

THEOREM 15.2. *Let E be an algebraic extension of finite degree over a subfield F. Then there exists a normal closure of E of finite degree over F. Further, if (N, ι) and (N', ι') are two normal closures of E over F there exists an isomorphism τ of N onto N' such that $\tau(\iota(a)) = \iota'(a)$ for all elements a of F.*

Proof. Let $\{x_1, \ldots, x_r\}$ be a basis for E over F and let

$m_1, \ldots, m_r$ be the minimum polynomials of $x_1, \ldots, x_r$ respectively over F. Let $f = m_1 m_2 \ldots m_r$.

(1) Let (N, ι) be any splitting field for f over E. Then N is also a splitting field for $\iota_P(f)$ over $\iota(F)$ and hence is a normal extension of $\iota(F)$. Suppose L is any subfield of N including $\iota(E)$ and normal over $\iota(F)$. Since each of the irreducible polynomials $\iota_P(m_i)$ has one root—namely $\iota(x_i)$—in L, it follows that each of these polynomials splits completely in $P(L)$, and so $L = N$. Thus (N, ι) is a normal closure of E over F and it is of course of finite degree over F.

(2) Suppose (N, ι) and (N', ι') are normal closures of E over F. Then $\iota_P(f)$ and $\iota'_P(f)$ split completely in $P(N)$ and $P(N')$ respectively; hence N and N' include splitting fields of these polynomials over $\iota(F)$ and $\iota'(F)$ respectively. But these splitting fields include $\iota(E)$, $\iota'(E)$ and are normal over $\iota(F)$, $\iota'(F)$ and hence must coincide with N, N' respectively (by condition (2) of the definition of normal closures). The existence of the isomorphism τ now follows from Theorem 11.2.

We remark that if E is a separable extension of F then the minimum polynomials $m_1, \ldots, m_r$ are all separable; hence, by the Corollary to Theorem 13.7, every splitting field of $f = m_1 m_2 \ldots m_r$ is a separable extension. In other words every normal closure of a separable extension is also separable.

From now on, if (N, ι) is a normal closure of an algebraic extension E of F we shall identify E and $\iota(E)$ and thus consider E and F as subfields of N. Suppose the normal closure N is a subfield of a field M and let τ be a monomorphism of E into M which acts like the identity on F (an F-monomorphism of E into M); we claim that $\tau(E)$ is included in N. To prove this assertion, let α be any element of E, $m = X^k + a_1 X^{k-1} + \ldots + a_k$ its minimum polynomial over F. Applying the monomorphism τ to the equation

$$\alpha^k + a_1\alpha^{k-1} + \ldots + a_k = 0,$$

we obtain

$$(\tau(\alpha))^k + a_1(\tau(\alpha))^{k-1} + \ldots + a_k = 0,$$

which shows that $\tau(\alpha)$ is also a root of m in M. But m is an irreducible polynomial in $P(F)$ with one root α in the normal extension N; thus m splits completely in $P(N)$ and so $\tau(\alpha)$ lies in N as required. Thus the importance of normal closures lies in the fact that, when we want to consider F-monomorphisms of E into fields including E, then once we have constructed the F-monomorphisms of E into a normal closure we obtain no further F-monomorphisms by considering larger extensions. We shall use this result to give a further characterisation of normal extensions of finite degree.

THEOREM 15.3. *Let E be an algebraic extension of finite degree over a subfield F. Then E is a normal extension of F if and only if every F-monomorphism of E into a normal extension of F including E is actually an F-automorphism of E itself.*

Proof. (1) Suppose E is a normal extension of F. Then E is itself a normal closure of E over F including E (and, in fact, the only one).

Let N be any normal extension of F including E, τ any F-monomorphism of E into N. According to the discussion preceding this theorem, $\tau(E)$ is included in the normal closure E. To show that τ is actually an F-automorphism of E we must prove that $\tau(E) = E$. From Theorem 7.1 we deduce that

$$(\tau(E): F) = (\tau(E): \tau(F)) = (E: F);$$

it now follows from Corollary 2 of Theorem 7.2 that $E = \tau(E)$ and hence τ is an F-automorphism of E as asserted.

(2) Conversely, let N be any normal extension of F

which includes E and suppose that every F-monomorphism of E into N is an automorphism of E.

Let f be any irreducible polynomial in $P(F)$ which has a root α_1 in E. Since N is a normal extension of F, f splits completely in $P(N)$; let α_2 be any other root of f in N. Then, according to Corollary 3 of Theorem 15.1, there exists an F-automorphism τ' of N such that $\tau'(\alpha_1) = \alpha_2$. Consider now the mapping τ of E into N defined by setting $\tau(x) = \tau'(x)$ for all elements x of E; τ is plainly an F-monomorphism of E into N. By hypothesis, τ is an automorphism of E; so $\tau(\alpha_1) = \tau'(\alpha_1) = \alpha_2$ belongs to E.

It follows that E is a normal extension of F, as required.

To illustrate the situation when E is not a normal extension of F we may refer to Example 3 of § 11. So suppose α is the unique real number such that $\alpha^3 = 2$ and set $E = \mathbf{Q}(\alpha)$; then E is not a normal extension of $\mathbf{Q}$ since the irreducible polynomial X^3-2 of $P(\mathbf{Q})$ has one root α in E but does not split completely in $P(E)$. The splitting field $K = \mathbf{Q}(\alpha, i\sqrt{3})$ which we constructed is easily seen to be a normal closure of E over $\mathbf{Q}$. There are three $\mathbf{Q}$-monomorphisms ε, τ, ρ of E into K, mapping α onto α, $\beta = \frac{1}{2}(-1+i\sqrt{3})\alpha$ and $\gamma = \frac{1}{2}(-1-i\sqrt{3})\alpha$ respectively. But only the first of these is an automorphism of E, for $\tau(E) = \mathbf{Q}(\beta)$ and $\rho(E) = \mathbf{Q}(\gamma)$ are not subfields of $\mathbf{R}$, while E itself is a subfield of $\mathbf{R}$.

Let E be an algebraic extension of a subfield F. We have already remarked that the maximum number of F-monomorphisms of E into a field K containing E is attained when we take K to be a normal closure of E over F. We shall now determine this maximum number in the case where E is a finite separable extension of F.

THEOREM 15.4. *Let E be a separable extension of finite degree n over a subfield F. Then there are precisely n distinct F-monomorphisms of E into a normal closure N*

of E over F (and hence into any normal extension of F including E).

Proof. We proceed by induction on the degree of E over F.

If $(E: F) = 1$, then $E = F$ and the result is obvious. For the only F-monomorphism of F itself into F is the identity automorphism of F.

Suppose now that we have established the result for all extensions of degree less than k and let E be an extension of degree k.

Let α be an element of E which does not belong to F. If m is the minimum polynomial of α over F, then $\partial m = (F(\alpha): F) = r$, say; of course $r \neq 1$. Now m is a separable irreducible polynomial in $P(F)$ which has one root α in the normal extension N of F; hence m splits completely in $P(N)$ and its roots $\alpha_1 = \alpha, \alpha_2, \ldots, \alpha_r$ in N are all distinct.

If we set $(E: F(\alpha)) = s$, then we have $rs = k$ and $1 \leqq s < k$. Since N is clearly a normal closure of E over $F(\alpha)$ we may apply the inductive hypothesis and deduce that there are precisely s monomorphisms $\rho_1, \ldots, \rho_s$ of E into N which act like the identity on $F(\alpha)$. Next, it follows from Corollary 3 of Theorem 15.1 that there are r distinct F-automorphisms of N, say $\tau_1, \ldots, \tau_r$, such that $\tau_i(\alpha) = \alpha_i$ $(i = 1, \ldots, r)$. Consider now the mappings ϕ_{ij} of E into N $(i = 1, \ldots, r;\ j = 1, \ldots, s)$ defined by setting $\phi_{ij}(x) = \tau_i(\rho_j(x))$ for all elements x of E. It is easily verified that these mappings are F-monomorphisms of E into N.

We claim that the $k = rs$ mappings ϕ_{ij} are all distinct. To this end we notice first that $\phi_{ij}(\alpha) = \tau_i(\rho_j(\alpha)) = \tau_i(\alpha) = \alpha_i$. Hence, if $\phi_{ij} = \phi_{lm}$, we deduce at once that $i = l$. If, next, $\phi_{ij} = \phi_{im}$, then for every element x of E we have $\tau_i(\rho_j(x)) = \tau_i(\rho_m(x))$, and hence, since τ_i is one-to-one, $\rho_j(x) = \rho_m(x)$ for every element x of E; that is to say

$\rho_j = \rho_m$, i.e., $j = m$. This discussion shows that there are at least k distinct F-monomorphisms of E into N.

To show that there are no more, we must prove that every F-monomorphism of E into N coincides with one of the mappings ϕ_{ij}. Let τ, then, be an F-monomorphism of E into N; τ must map α onto another root of m in N, say $\tau(\alpha) = \alpha_i$. Consider the mapping ϕ of E into N defined by setting $\phi(x) = \tau_i^{-1}(\tau(x))$ for all elements x of E; this is clearly an $F(\alpha)$-monomorphism of E into N and hence coincides with one of the mappings $\rho_1, \ldots, \rho_s$; say $\phi = \rho_j$. Then for every element x of E we have

$$\phi_{ij}(x) = \tau_i(\rho_j(x)) = \tau_i(\phi(x)) = \tau(x),$$

and so $\tau = \phi_{ij}$, as required.

This completes the induction.

COROLLARY. *Let K be a separable normal extension of finite degree n over a subfield F. Then there are exactly n distinct F-automorphisms of K.*

Proof. According to the theorem, there are n distinct F-monomorphisms of K into a normal closure of K including K; included among these are all the F-automorphisms of K. But, by Theorem 15.3, all the F-monomorphisms are in fact F-automorphisms. Hence the result follows.

If E is a separable extension of finite degree n over a subfield F and $\tau_1, \ldots, \tau_n$ are the n distinct F-monomorphisms of E into a normal extension N of F including E, the subfields $\tau_1(E), \ldots, \tau_n(E)$ of N are called the **conjugates** of E over F in N. Although the monomorphisms $\tau_1, \ldots, \tau_n$ are all distinct, the conjugates may not be distinct; indeed, if E is itself a normal extension of F, it follows from Theorem 15.3 that its conjugates all coincide with E itself.

We conclude this section with an important characterisation of separable normal extensions of finite degree.

We remark first that if E is any extension of a subfield F, then the set G of F-automorphisms of E is a subgroup of the group A of all automorphisms of E. For, if τ and ρ are F-automorphisms of E and a is any element of F we have $(\tau\rho)(a) = \tau(\rho(a)) = \tau(a) = a$, and further, since $\tau(a) = a$, it follows on applying τ^{-1} to both sides of this relation that $a = \tau^{-1}(\tau(a)) = \tau^{-1}(a)$; thus $\tau\rho$ and τ^{-1} belong to G and so G is a subgroup of A.

The field F is certainly included in the fixed field F_0 under the subgroup G; but one must guard against supposing that F and F_0 must be the same. For example, let us consider again Example 3 of § 11, in which α is the unique real number such that $\alpha^3 = 2$. Then there is only one $\mathbf{Q}$-automorphism of $E = \mathbf{Q}(\alpha)$, namely the identity automorphism ε. But the fixed field under the group of automorphisms consisting of ε alone is the whole field E. Of course, as we have already noticed, $\mathbf{Q}(\alpha)$ is not a normal extension of $\mathbf{Q}$.

THEOREM 15.5. *Let K be a finite extension of a subfield F, G the group of F-automorphisms of K, and F_0 the fixed field under G. Then K is a separable normal extension of F if and only if $F = F_0$.*

Proof. (1) Suppose K is a separable normal extension of F; let $(K\colon F) = n$.

According to the Corollary to Theorem 15.4, the group G has order n. It now follows from Theorem 14.2 that $(K\colon F_0) = n$. Hence, since F is included in F_0, we have $(F_0\colon F) = 1$ and so $F = F_0$.

(2) Conversely, suppose F is the fixed field under the group G of F-automorphisms of K; let

$$G = \{\tau_1 = \varepsilon, \tau_2, \ldots, \tau_n\}.$$

Let f be an irreducible polynomial in $P(F)$ which has a root α in K. The images of α under the n automorphisms $\tau_1, \ldots, \tau_n$ may not all be distinct. By renumbering the automorphisms if necessary, we may suppose that $\alpha_1 = \alpha$, $\alpha_2 = \tau_2(\alpha)$, ..., $\alpha_r = \tau_r(\alpha)$ are the distinct images of α under the automorphisms. Then for each automorphism τ_i in G, the set $\{\tau_i(\alpha_1), ..., \tau_i(\alpha_r)\}$ coincides with $\{\alpha_1, ..., \alpha_r\}$ —that is to say, the automorphisms in G simply permute the elements $\alpha_1, ..., \alpha_r$ among themselves. (To see this, we have only to remark that $\tau_i(\alpha_j) = \tau_i(\tau_j(\alpha)) = (\tau_i\tau_j)(\alpha)$, and hence, since $\tau_i\tau_j$ belongs to G, $\tau_i(\alpha_j)$ is one of the images of α.) It follows that if e_k is the kth elementary symmetric function of $\alpha_1, ..., \alpha_r$, i.e., the sum of the products of these elements taken k at a time ($k = 1, ..., r$), we have $\tau_i(e_k) = e_k$ for all the automorphisms τ_i, and hence e_k belongs to $F_0 = F$.

Consider now the polynomial

$$g = X^r - e_1X^{r-1} + e_2X^{r-2} - ... + (-1)^r e_r$$

in $P(F)$; this polynomial splits completely in $P(K)$:

$$g = (X-\alpha_1)(X-\alpha_2)...(X-\alpha_r).$$

We claim that g is the minimum polynomial of $\alpha_1 = \alpha$ over F. So suppose that

$$h = a_0 + a_1X + ... + a_sX^s$$

is a polynomial in $P(F)$ which has α as a root. Thus we have

$$a_0 + a_1\alpha + ... + a_s\alpha^s = 0.$$

Applying the automorphisms $\tau_1, ..., \tau_r$ to this relation we obtain

$$a_0 + a_1\alpha_i + ... + a_s\alpha_i^s = 0 \quad (i = 1, ..., r);$$

so $\alpha_1, ..., \alpha_r$ are roots of h and hence g is a factor of h. Thus g is a monic polynomial of least degree among those polynomials in $P(F)$ which have α as a root, i.e., g is the minimum polynomial of α over F.

It follows that the original irreducible polynomial f is simply a constant multiple of g; hence f splits completely in $P(K)$, and all its roots are distinct. Thus K is a separable normal extension of F, as asserted.

§ **16. The fundamental theorem of Galois theory.** Let K be an extension of a subfield F. The group G of F-automorphisms of K is called the **Galois group** of K over F. Let $\mathscr{K}$ be the set of subfields of K which include F, and let $\mathscr{G}$ be the set of subgroups of the Galois group G. In the case where K is a separable normal extension of finite degree over F, the fundamental theorem of Galois theory establishes a one-to-one correspondence between $\mathscr{K}$ and $\mathscr{G}$ which enables us to deduce information about the subfields of K which include F from a knowledge of the Galois group of K over F.

We define a mapping Γ of $\mathscr{K}$ into $\mathscr{G}$ and a mapping Φ of $\mathscr{G}$ into $\mathscr{K}$ as follows. For every subfield E of K including F, let $\Gamma(E)$ be the subgroup of G consisting of those F-automorphisms of K which act like the identity on E, i.e., of the E-automorphisms of K; in other words, $\Gamma(E)$ is the Galois group of K over E. For every subgroup H of G we define $\Phi(H)$ to be the fixed field under H; $\Phi(H)$ is of course a subfield of K including F. We remark at once that the mappings Γ and Φ " reverse inclusions "; that is to say, if E_1 is included in E_2 then $\Gamma(E_2)$ is included in $\Gamma(E_1)$, and if H_1 is included in H_2 then $\Phi(H_2)$ is included in $\Phi(H_1)$. Suppose, for example, that E_1 is included in E_2 and γ belongs to $\Gamma(E_2)$; then $\gamma(x) = x$ for all elements x of E_2, and so, in particular, $\gamma(t) = t$ for all elements t of E_1; hence γ belongs to $\Gamma(E_1)$. The other result is established by a similar simple argument.

THEOREM 16.1. (*Fundamental Theorem, Part* 1). *Let K be a separable normal extension of finite degree n over a subfield F, with Galois group G. Then G has order n. Let $\mathscr{K}$ be the set of subfields of K which include F, $\mathscr{G}$ the set of subgroups of G. Then the mapping Γ defined above is a one-to-one* *mapping of $\mathscr{K}$ onto $\mathscr{G}$. Further, for every subfield E in K, the degree of K over E is equal to the order of $\Gamma(E)$ and the degree of E over F is equal to the index of $\Gamma(E)$ in G.*

Proof. The fact that G has order n has already been established in the Corollary to Theorem 15.4.

Let E be any subfield of K including F. According to Corollary 1 of Theorem 15.1, K is a normal extension of E; since K is a separable extension of F it is also a separable extension of E (Theorem 13.4). Thus it follows from the Corollary to Theorem 15.4 that $(K: E)$ is equal to the order of $\Gamma(E)$. Then we have $(E: F) = (K: F)/(K: E)$ $=$ (order of G)/(order of $\Gamma(E)$) $=$ index of $\Gamma(E)$ in G.

Now, since K is a separable normal extension of E, it follows from Theorem 15.5 that E is the precise fixed field under the group of automorphisms of K which act like the identity of E; that is to say, $E = \Phi(\Gamma(E))$. It now follows at once that Γ is a one-to-one mapping. For if E_1 and E_2 are subfields of K including F such that $\Gamma(E_1) = \Gamma(E_2)$, it follows that $\Phi(\Gamma(E_1)) = \Phi(\Gamma(E_2))$ and hence that $E_1 = E_2$.

Next, let H be any subgroup of G. It is clear that H is included in $H_1 = \Gamma(\Phi(H))$. Further, if we take $E = \Phi(H)$ in the preceding paragraph, we have

$$\Phi(H) = \Phi(\Gamma(\Phi(H))) = \Phi(H_1).$$

Since H and H_1 are finite groups it follows from Theorem 14.2 that the order of $H = (K: \Phi(H)) = (K: \Phi(H_1)) =$ the order of H_1. Thus $H = H_1 = \Gamma(\Phi(H))$, and hence Γ maps $\mathscr{K}$ onto $\mathscr{G}$.

This completes the proof.

If E_1 and E_2 are subfields of K including F which are isomorphic to one another under an isomorphism which acts like the identity on F it is natural to expect that the corresponding subgroups $\Gamma(E_1)$ and $\Gamma(E_2)$ of G are isomorphic. We show next that this is in fact the case.

THEOREM 16.2. *Let K be a separable normal extension of finite degree over a subfield F. Let E be a subfield of K including F, τ an F-automorphism of K. Then*

$$\Gamma(\tau(E)) = \tau\Gamma(E)\tau^{-1}.$$

Proof. Let us write $E_1 = \tau(E)$, $H = \Gamma(E)$, $H_1 = \Gamma(E_1)$. Then we have to prove that $H_1 = \tau H\tau^{-1}$.

First let γ be any element of H; we shall show that $\tau\gamma\tau^{-1}$ belongs to H_1. So let x_1 be any element of E_1. Then there is an element x of E such that $x_1 = \tau(x)$ and we have

$$\begin{aligned}(\tau\gamma\tau^{-1})(x_1) = (\tau\gamma\tau^{-1})(\tau(x)) &= (\tau\gamma)((\tau^{-1}\tau)(x)) \\ &= (\tau\gamma)(x) = \tau(\gamma(x)) = \tau(x) = x_1.\end{aligned}$$

Thus $\tau\gamma\tau^{-1}$ belongs to H_1 and hence $\tau H\tau^{-1}$ is included in H_1.

Conversely, let γ_1 be any element of H_1; we shall prove that $\gamma = \tau^{-1}\gamma_1\tau$ belongs to H—it will then follow that $\gamma_1 = \tau\gamma\tau^{-1}$ belongs to $\tau H\tau^{-1}$. So we let x be any element of E. Since $\tau(x)$ is an element of E_1, it is left fixed by γ_1; thus $(\gamma_1\tau)(x) = \tau(x)$ and so

$$\gamma(x) = \tau^{-1}((\gamma_1\tau)(x)) = \tau^{-1}(\tau(x)) = x.$$

Hence γ belongs to H, as asserted, and γ_1 belongs to $\tau H\tau^{-1}$. Thus H_1 is included in $\tau H\tau^{-1}$.

It follows that $H_1 = \tau H\tau^{-1}$ as required.

In the terminology which we introduced after Theorem 15.4, E_1 is one of the conjugates of E over F in K. Theorem 16.2 shows that the subgroup H_1 of G corresponding to E_1 is conjugate to the subgroup corresponding to E. We have already remarked that if E is a normal extension of F then it coincides with all its conjugates. Now a subgroup which coincides with all its conjugate subgroups is called a normal subgroup. It is thus natural to expect that if a subfield E of K is normal over F then the subgroup of G corresponding to it is normal and conversely. The second part of the Fundamental Theorem shows that this is in fact the case and also gives information about the Galois group of E over F.

THEOREM 16.3. (*Fundamental Theorem, Part* 2). *Let K be a separable normal extension of finite degree over a subfield F, with Galois group G. A subfield L of K including F is a normal extension of F if and only if the subgroup $\Gamma(L)$ of G corresponding to L is a normal subgroup of G. If L is a normal extension of F the Galois group of L over F is isomorphic to the factor group $G/\Gamma(L)$.*

Proof. Let us write $N = \Gamma(L)$.

(1) Suppose L is a normal extension of F.

Let τ be any element of G. Define the mapping τ' of L into K by setting $\tau'(x) = \tau(x)$ for every element x of L. Since τ is an F-automorphism of K it is easy to see that τ' is an F-monomorphism of L into K, and hence an F-automorphism of L (by Theorem 15.3). Thus

$$\tau(L) = \tau'(L) = L.$$

It follows from Theorem 16.2 that $\tau N \tau^{-1} = N$; so N is a normal subgroup of G, as asserted.

Let G' be the Galois group of L over F. Consider the mapping ϕ of G into G' defined by setting $\phi(\tau) = \tau'$ (as described above) for every element τ of G. This mapping ϕ is a group homomorphism (see § 3) of G into G'; for, if τ_1 and τ_2 are elements of G and x is an element of L,

$$(\phi(\tau_1\tau_2))(x) = (\tau_1\tau_2)(x) = \tau_1(\tau_2(x))$$

and

$$(\phi(\tau_1)\phi(\tau_2))(x) = \phi(\tau_1)(\phi(\tau_2)(x)) = \tau_1(\tau_2(x)).$$

So $\phi(\tau_1\tau_2) = \phi(\tau_1)\phi(\tau_2)$. Corollary 2 to Theorem 15.1 shows that ϕ actually maps G onto G'. Now the kernel of the homomorphism ϕ (i.e., the set of elements τ in G such that $\phi(\tau)$ is the identity of G') consists of those F-automorphisms of K which act like the identity on L; so the kernel is the subgroup $\Gamma(L) = N$. It now follows that G' is isomorphic to the factor group G/N.

(2) Suppose conversely that N is a normal subgroup of G.

Let τ_1 be any F-monomorphism of L into K. According to Corollary 2 of Theorem 15.1, there is an F-automorphism τ of K which acts like τ_1 on L. Since N is a normal subgroup of G, we have $\tau N \tau^{-1} = N$ and hence, by Theorem 16.2, $\Gamma(\tau(L)) = \Gamma(L)$. Since Γ is a one-to-one mapping, it follows that $\tau_1(L) = \tau(L) = L$. Thus τ_1 is actually an F-automorphism of L. Hence, by Theorem 15.3, L is a normal extension of F, as asserted.

COROLLARY. *Let K be a separable normal extension of finite degree over a subfield F, with Galois group G. Let $E_0 = F, E_1, \ldots, E_r = K$ be a sequence of subfields of K, each included in the next, and let $H_i = \Gamma(E_i)$ $(i = 0, 1, \ldots, r)$. If E_i is a normal extension of E_{i-1} then H_i is a normal subgroup of H_{i-1} and the Galois group of E_i over E_{i-1} is isomorphic to the factor group H_{i-1}/H_i. Conversely, if H_i is a normal subgroup of H_{i-1}, then E_i is a normal extension of E_{i-1}.*

Before giving examples to illustrate the results of this section we mention a simple result concerning the Galois groups of isomorphic extensions.

THEOREM 16.4. *Let K, K' be separable normal extensions of finite degree over their subfields F, F' with Galois groups G, G' respectively. If there is an isomorphism σ of K into K' such that $\sigma(F) = F'$, then $G'$and G' are isomorphic groups.*

Proof. We define a mapping ϕ of G into the set of mappings of $K' = \sigma(K)$ into itself as follows. For each element τ of G we define $\phi(\tau)$ to be the mapping of K' into itself given by

$$[\phi(\tau)](\sigma(x)) = \sigma(\tau(x))$$

for all elements x in K. It is then a routine matter to verify that $\phi(\tau)$ is actually an F'-automorphism of K', i.e., that $\phi(\tau)$ is in G', and then that ϕ is an isomorphism of G onto G'.

Example 1. Let $F = \mathbf{Q}$ and let K be the subfield $\mathbf{Q}(i\sqrt{3})$ of $\mathbf{C}$. We saw in Example 2 of § 11 that K is a splitting field over F of the polynomial $X^2 - X + 1$. Hence K is a normal extension of F (Theorem 15.1) and K is a separable extension of F since fields of characteristic zero have no inseparable extensions (Theorem 13.3). We proved also that $(K\colon F) = 2$; so the Galois group of K over F is of order 2. Since $\{1, i\sqrt{3}\}$ is a basis for K over F, every element x of K can be expressed uniquely in the form $x = a + bi\sqrt{3}$ where a and b are rational numbers. Since $i\sqrt{3}$ is a root of the polynomial $X^2 + 3$ in $P(F)$, any F-automorphism of K must map $i\sqrt{3}$ onto a root of this polynomial in K, and so onto either $i\sqrt{3}$ or $-i\sqrt{3}$. Thus the two elements of the Galois group are the F-automorphisms ε and τ given by

$$\varepsilon(a + bi\sqrt{3}) = a + bi\sqrt{3} \text{ and } \tau(a + bi\sqrt{3}) = a - bi\sqrt{3}.$$

Clearly $\tau^2 = \varepsilon$ and the Galois group is cyclic of order 2. Since a cyclic group of order 2 has no subgroups except itself and the subgroup consisting of the identity alone, it follows from the results of this section that the only subfields of K which include F are K itself (corresponding to the subgroup $\{\varepsilon\}$) and F (corresponding to the whole group).

Example 2. Let $F = \mathbf{Q}$ and let K be the subfield $\mathbf{Q}(\alpha, i)$ of $\mathbf{C}$, where α is the positive real number such that

$\alpha^4 = 2$. We saw in Example 4 of § 11 that K is a splitting field over F of the polynomial $X^4 - 2$ and hence is a separable normal extension of F. We proved that $(K: F) = 8$; so the Galois group G of K over F is of order 8.

In order to determine this Galois group we remark first that, since $\{1, \alpha, \alpha^2, \alpha^3\}$ is a basis over F for $F(\alpha)$ and $\{1, i\}$ is a basis over $F(\alpha)$ for K, the proof of Theorem 7.2 shows that $\{1, \alpha, \alpha^2, \alpha^3, i, i\alpha, i\alpha^2, i\alpha^3\}$ is a basis for K over F. Hence every element x of K can be expressed uniquely in the form

$$x = a_1 + a_2\alpha + a_3\alpha^2 + a_4\alpha^3 + a_5 i + a_6 i\alpha + a_7 i\alpha^2 + a_8 i\alpha^3, \quad (16.1)$$

where $a_1, \ldots, a_8$ are rational numbers. It follows that, once we know how an F-automorphism of K acts on α and i, we know how it acts on every element of K.

Since α is a root of the polynomial $X^4 - 2$ in $P(F)$, each F-automorphism must map α onto one of the roots of this polynomial in K; similarly, since i is a root of the polynomial $X^2 + 1$ in $P(F)$, each F-automorphism of K must map i onto one of the roots of this polynomial in K. These remarks allow us to define the eight elements of G by means of the table

	τ_1	τ_2	τ_3	τ_4	τ_5	τ_6	τ_7	τ_8
α	α	$i\alpha$	$-\alpha$	$-i\alpha$	α	$i\alpha$	$-\alpha$	$-i\alpha$
i	i	i	i	i	$-i$	$-i$	$-i$	$-i$

which describes their action on α and i.

Of course τ_1 is the identity automorphism ε. Now set $\tau_2 = \sigma$; we claim that $\tau_3 = \sigma^2$, $\tau_4 = \sigma^3$ and $\sigma^4 = \varepsilon$. To prove these results we examine the effects of the various automorphisms on α and i; for example,

$$\sigma^2(\alpha) = \sigma(\sigma(\alpha)) = \sigma(i\alpha) = \sigma(i)\sigma(\alpha) = i^2\alpha = -\alpha = \tau_3(\alpha)$$

and $$\sigma^2(i) = \sigma(\sigma(i)) = \sigma(i) = i = \tau_3(i),$$

whence $\tau_3 = \sigma^2$. The other results follow by similar computations. Next set $\tau_5 = \tau$; then $\tau^2 = \varepsilon$, $\tau_6 = \sigma\tau$, $\tau_7 = \sigma^2\tau$ and $\tau_8 = \sigma^3\tau$. Finally we have $\tau\sigma = \sigma^3\tau$: for

$$(\tau\sigma)(\alpha) = \tau(\sigma(\alpha)) = \tau(i\alpha) = \tau(i)\tau(\alpha) = -i\alpha = (\sigma^3\tau)(\alpha)$$

and $$(\tau\sigma)(i) = \tau(\sigma(i)) = \tau(i) = -i = (\sigma^3\tau)(i).$$

This discussion shows that G is isomorphic to the dihedral group of order 8.

The group G has four normal subgroups other than itself and the identity subgroup—namely $N_1 = \{\varepsilon, \sigma, \sigma^2, \sigma^3\}$, $N_2 = \{\varepsilon, \sigma\tau, \sigma^2, \sigma^3\tau\}$, $N_3 = \{\varepsilon, \tau, \sigma^2, \sigma^2\tau\}$ and $N_4 = \{\varepsilon, \sigma^2\}$ —and also four non-normal subgroups—$H_1 = \{\varepsilon, \tau\}$, $H_2 = \{\varepsilon, \sigma\tau\}$, $H_3 = \{\varepsilon, \sigma^2\tau\}$ and $H_4 = \{\varepsilon, \sigma^3\tau\}$. The inclusion relations between these subgroups are shown in the diagram in fig. 5, where all the arrows indicate the appropriate inclusion monomorphisms.

In order to describe the fixed fields under these subgroups we examine the effect of the eight automorphisms on a typical element x of K (see (16.1)). We obtain

$$\begin{aligned}
\varepsilon(x) &= a_1+a_2\alpha+a_3\alpha^2+a_4\alpha^3+a_5i+a_6i\alpha+a_7i\alpha^2+a_8i\alpha^3\\
\sigma(x) &= a_1-a_6\alpha-a_3\alpha^2+a_8\alpha^3+a_5i+a_2i\alpha-a_7i\alpha^2-a_4i\alpha^3\\
\sigma^2(x) &= a_1-a_2\alpha+a_3\alpha^2-a_4\alpha^3+a_5i-a_6i\alpha+a_7i\alpha^2-a_8i\alpha^3\\
\sigma^3(x) &= a_1+a_6\alpha-a_3\alpha^2-a_8\alpha^3+a_5i-a_2i\alpha-a_7i\alpha^2+a_4i\alpha^3\\
\tau(x) &= a_1+a_2\alpha+a_3\alpha^2+a_4\alpha^3-a_5i-a_6i\alpha-a_7i\alpha^2-a_8i\alpha^3\\
(\sigma\tau)(x) &= a_1+a_6\alpha-a_3\alpha^2-a_8\alpha^3-a_5i+a_2i\alpha+a_7i\alpha^2-a_4i\alpha^3\\
(\sigma^2\tau)(x) &= a_1-a_2\alpha+a_3\alpha^2-a_4\alpha^3-a_5i+a_6i\alpha-a_7i\alpha^2+a_8i\alpha^3\\
(\sigma^3\tau)(x) &= a_1-a_6\alpha-a_3\alpha^2+a_8\alpha^3-a_5i-a_2i\alpha+a_7i\alpha^2+a_4i\alpha^3.
\end{aligned}$$

The element x belongs to the fixed field $\Phi(N_1)$ under N_1 if and only if $x = \sigma(x) = \sigma^2(x) = \sigma^3(x)$. Comparing the coefficients of the basis elements, we see that these equations hold if and only if $a_2 = -a_2$, $a_3 = -a_3$, $a_4 = -a_4$, $a_6 = -a_6$, $a_7 = -a_7$, $a_8 = -a_8$, i.e., if and only if $a_2 = a_3 = a_4 = a_6 = a_7 = a_8 = 0$. Thus the elements of $\Phi(N_1)$ are precisely those which can be expressed in the form $a_1 + a_5 i$, where a_1 and a_5 are rational numbers;

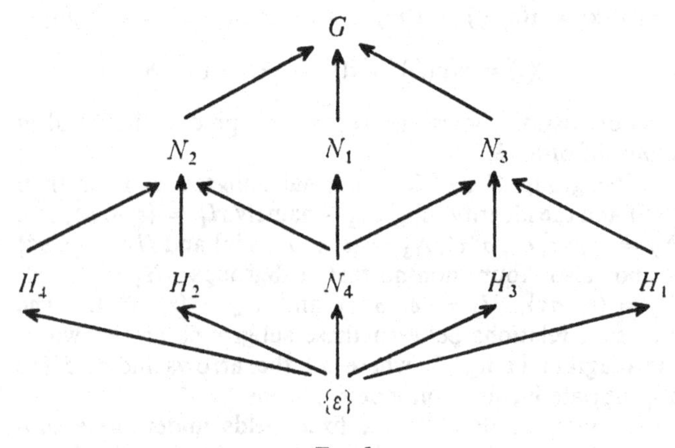

FIG. 5

in other words, $\Phi(N_1) = F(i)$. Since $F(i)$ is a splitting field over F for the polynomial X^2+1, it is a normal extension of F, as it should be in accordance with Theorem 16.3. The Galois group of $F(i)$ over F is cyclic of order 2, as is the factor group G/N_1.

Similar arguments show that $\Phi(N_2) = F(i\alpha^2) = F(i\sqrt{2})$ and $\Phi(N_3) = F(\alpha^2) = F(\sqrt{2})$. These fields are also normal extensions of F, being splitting fields over F for X^2+2 and X^2-2 respectively; the Galois group in each case is cyclic of order 2.

Next we notice that x belongs to the fixed field $\Phi(N_4)$

under N_4 if and only if $x = \sigma^2(x)$, i.e., if and only if $a_2 = -a_2$, $a_4 = -a_4$, $a_6 = -a_6$, $a_8 = -a_8$, i.e., if and only if $a_2 = a_4 = a_6 = a_8 = 0$. So the elements of $\Phi(N_4)$ are precisely those which can be expressed in the form $a_1 + a_3\alpha^2 + a_5 i + a_7 i\alpha^2$. Thus $\Phi(N_4) = F(i, \alpha^2) = F(i, \sqrt{2})$; this is a normal extension of F, being a splitting field for $(X^2+1)(X^2-2)$ over F. It is not hard to verify that the Galois group of $F(i, \sqrt{2})$ over F is isomorphic to the Klein

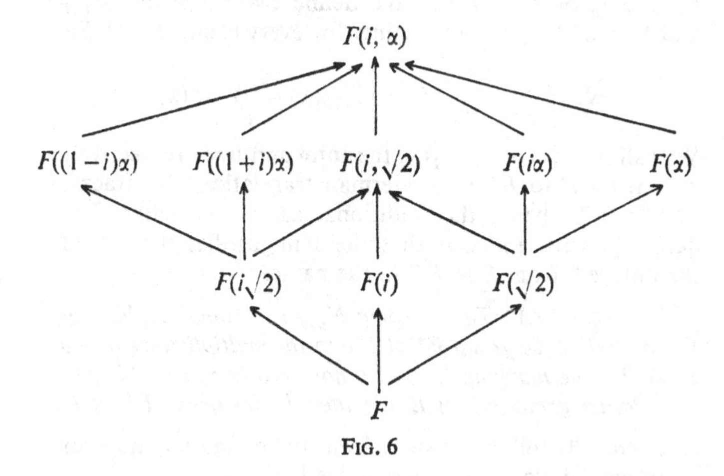

FIG. 6

four-group, and that G/N_4 is also isomorphic to this group.

Similar investigations show that $\Phi(H_1) = F(\alpha)$, $\Phi(H_2) = F((1+i)\alpha)$, $\Phi(H_3) = F(i\alpha)$ and $\Phi(H_4) = F((1-i)\alpha)$. As an example we consider $\Phi(H_2)$: x is left fixed by H_2 if and only if $x = (\sigma\tau)(x)$; this occurs if and only if $a_2 = a_6$, $a_3 = -a_3$, $a_4 = -a_8$, $a_5 = -a_5$. Thus x belongs to $\Phi(H_2)$ if and only if it has the form

$$a_1 + a_2(1+i)\alpha + a_4(1-i)\alpha^3 + a_7 i\alpha^2 = a_1 + a_2(1+i)\alpha + \tfrac{1}{2}a_7((1+i)\alpha)^2 - \tfrac{1}{2}a_4((1+i)\alpha)^3.$$

The inclusion relations between the subfields of K are shown in fig. 6, where as usual all the arrows indicate inclusion monomorphisms.

§ **17. Norms and traces.** Let E be a separable extension of finite degree n over a subfield F. Let K be a normal closure of E over F which contains E. Then, according to Theorem 15.4, there are exactly n distinct F-monomorphisms $\tau_1, \ldots, \tau_n$ of E into K. We define two mappings $N_{E/F}$ and $S_{E/F}$ of E into K by setting, for every element x of E,

$$N_{E/F}(x) = \prod_{i=1}^{n} \tau_i(x), \quad S_{E/F}(x) = \sum_{i=1}^{n} \tau_i(x).$$

We call $N_{E/F}(x)$ and $S_{E/F}(x)$ the **norm** and **trace** respectively of x from E to F. (The German translation of " trace " is " Spur ": hence the traditional use of the letter S to denote the trace.) Our first theorem justifies the use of the phrase " from E to F " in the names.

THEOREM 17.1. *The mapping $N_{E/F}$ is a homomorphism of the multiplicative group E^* of E into the multiplicative group F^* of F; the mapping $S_{E/F}$ is a non-zero homomorphism of the additive group E^+ of E into the additive group F^+ of F.*

Proof. It follows at once from the definitions that for every pair of elements x, y of E we have

$$N_{E/F}(xy) = \prod \tau_i(xy) = \prod \tau_i(x)\tau_i(y) = (\prod \tau_i(x))(\prod \tau_i(y)) = N_{E/F}(x)N_{E/F}(y)$$

and similarly

$$S_{E/F}(x+y) = S_{E/F}(x) + S_{E/F}(y).$$

This shows that $N_{E/F}$ and $S_{E/F}$ are homomorphisms of E^* and E^+ into K^* and K^+ respectively.

Let now τ be any F-automorphism of K. The mappings $\rho_1, \ldots, \rho_n$ of E into K defined by setting $\rho_i(x) = \tau(\tau_i(x))$ for every element x of E ($i = 1, \ldots, n$) are clearly n distinct

F-monomorphisms of E into K. Thus the set $\{\rho_1, ..., \rho_n\}$ coincides with the set $\{\tau_1, ..., \tau_n\}$ (apart from order, of course). Let x be any element of E; then we have

$$\tau(N_{E/F}(x)) = \tau\left(\prod_{i=1}^{n} \tau_i(x)\right) = \prod_{i=1}^{n} \tau(\tau_i(x))$$

$$= \prod_{i=1}^{n} \rho_i(x) = N_{E/F}(x)$$

and similarly $\tau(S_{E/F}(x)) = S_{E/F}(x)$. So the norm and trace of x belong to the fixed field under all the F-automorphisms of K. Since K is a normal closure of a separable extension it is a separable normal extension of finite degree over F (see the remark just after Theorem 15.2). Hence it follows from Theorem 15.5 that the fixed field under all the F-automorphisms of K is just F itself. Hence $N_{E/F}(x)$ and $S_{E/F}(x)$ belong to F as required.

All that remains is to show that $S_{E/F}$ is not the zero homomorphism. Suppose, to the contrary, that

$$S_{E/F}(x) = \tau_1(x) + ... + \tau_n(x) = 0$$

for all elements x of E; this would imply, however, that the set $\{\tau_1, ..., \tau_n\}$ of distinct monomorphisms of E into K is linearly dependent over K. Since this contradicts Theorem 14.1, the desired result follows.

Example 1. Let $F = \mathbf{Q}$, $E = \mathbf{Q}(i)$. Then E is a normal extension of F and the two F-automorphisms τ_1, τ_2 of E are given by

$$\tau_1(a+bi) = a+bi \text{ and } \tau_2(a+bi) = a-bi$$

for all elements $a+bi$ in E (a and b are rational numbers). Then we have

$$N_{E/F}(a+bi) = (a+bi)(a-bi) = a^2+b^2,$$

$$S_{E/F}(a+bi) = (a+bi)+(a-bi) = 2a.$$

Example 2. Let $F = \mathbf{Q}$ and let $E = \mathbf{Q}(\alpha)$ where α is the unique positive real number such that $\alpha^3 = 2$. Let $\beta = \frac{1}{2}(-1+i\sqrt{3})\alpha$ and $\gamma = \frac{1}{2}(-1-i\sqrt{3})\alpha$ be the remaining roots of X^3-2 in $\mathbf{C}$ (see Example 1, § 10). The normal closure of E over F in $\mathbf{C}$ is the splitting field $K = \mathbf{Q}(\alpha, i\sqrt{3})$ of X^3-2 (see Example 3, § 11). The three F-monomorphisms τ_1, τ_2, τ_3 of E into K are defined by setting

$$\tau_1(\alpha) = \alpha,\ \tau_2(\alpha) = \beta,\ \tau_3(\alpha) = \gamma,$$

$$\tau_i(a) = a \text{ for all elements } a \text{ of } F\ (i = 1, 2, 3).$$

Write $\omega = \frac{1}{2}(-1+i\sqrt{3})$; then $\omega^2 = \frac{1}{2}(-1-i\sqrt{3})$, $\omega^3 = 1$ and $1+\omega+\omega^2 = 0$. Let now $x = a+b\alpha+c\alpha^2$ be any element of E, where a, b, c are elements of F. Then we easily compute

$$\begin{aligned} N_{E/F}(x) &= \tau_1(x)\tau_2(x)\tau_3(x) \\ &= (a+b\alpha+c\alpha^2)(a+b\omega\alpha+c\omega^2\alpha^2)(a+b\omega^2\alpha+c\omega\alpha^2) \\ &= a^3+2b^3+4c^3-6abc, \\ S_{E/F}(x) &= (a+b\alpha+c\alpha^2)+(a+b\omega\alpha+c\omega^2\alpha^2) \\ &\qquad\qquad +(a+b\omega^2\alpha+c\omega\alpha^2) \\ &= 3a. \end{aligned}$$

Let now D be a separable extension of finite degree over a subfield F; Let E be a subfield of D which includes F. Then D is a separable extension of E and E is a separable extension of F (Theorem 13.4). Thus if x is any element of D we may form the norm of x from D to E, which is an element of E by Theorem 17.1, and form the norm of this element from E to F, so obtaining an element of F; we may also, of course, form the norm of x from D to F. The next theorem shows that these two procedures lead to the same element of F.

THEOREM 17.2. *Let D be a separable extension of finite degree over a subfield F; let E be a subfield of D including F. Then, for every element x of D, $N_{E/F}(N_{D/E}(x)) = N_{D/F}(x)$ and $S_{E/F}(S_{D/E}(x)) = S_{D/F}(x)$.*

Proof. Let K be a normal closure of D over F including D. Let $(E: F) = n$, $(D: E) = m$; then $(D: F) = mn$. There are thus n distinct F-monomorphisms $\sigma_1, \ldots, \sigma_n$ of E into K and m distinct E-monomorphisms $\tau_1, \ldots, \tau_m$ of D into K. According to Corollary 2 of Theorem 15.1 there are n distinct F-automorphisms $\sigma'_1, \ldots, \sigma'_n$ of K which act like $\sigma_1, \ldots, \sigma_n$ respectively on E. Let ϕ_{ij} ($i = 1, \ldots, n$; $j = 1, \ldots, m$) be the mappings of D into K defined by setting $\phi_{ij}(x) = \sigma'_i(\tau_j(x))$ for all elements x of D. These mn mappings are distinct F-monomorphisms of D into K (cf. the proof of Theorem 15.4) and hence they form a complete set of F-monomorphisms of D into K.

Let now x be any element of D. We have

$$\begin{aligned} N_{D/F}(x) &= \prod_{i,j} \phi_{ij}(x) \\ &= \prod_{i,j} \sigma'_i(\tau_j(x)) = \prod_i \sigma'_i\Big(\prod_j \tau_j(x)\Big) = \prod_i \sigma'_i\,(N_{D/E}(x)) \\ &= \prod_i \sigma_i(N_{D/E}(x)) = N_{E/F}(N_{D/E}(x)) \text{ as required.} \end{aligned}$$

The result for the traces follows similarly.

§ 18. The primitive element theorem and Lagrange's theorem. Let E be an extension of a subfield F. If there exists an element α of E such that $E = F(\alpha)$—so that E is a simple extension of F—we call α a **primitive element** for E over F. Our next theorem shows that for every extension of the type we are most concerned with—separable extensions of finite degree—there is always a primitive element.

THEOREM 18.1 (*Primitive element theorem*). *Let E be a separable extension of finite degree over a subfield F. Then there exists a primitive element for E over F.*

Proof. When F is a finite field this result will follow from our general discussion of finite fields in § 20. So we shall suppose for the remainder of the proof that F has infinitely many elements.

First we show that there are only finitely many subfields of E including F. To this end we let K be a normal closure of E over F which includes E; K is a normal separable extension of finite degree over F, with Galois group G say. Now the number of subfields of E including F is at most equal to the number of subfields of K including F; according to Theorem 16.1, however, this is equal to the number of subgroups of the (finite) group G, and hence is finite.

Let α_0 be an element of E such that, for every element α of E, $(F(\alpha): F) \leqq (F(\alpha_0): F)$; such an element clearly exists, since all the degrees $(F(\alpha): F)$ are less than or equal to $(E: F)$. We claim that α_0 is a primitive element for E over F, i.e., that $E = F(\alpha_0)$. Suppose, to the contrary, that $E \neq F(\alpha_0)$; then there exists an element β of E which does not belong to $F(\alpha_0)$, so that $(F(\alpha_0, \beta): F) > (F(\alpha_0): F)$. For each element a of F, $F(\alpha_0, \beta)$ includes the simple extension $F(\alpha_0 + a\beta)$. Since there are infinitely many elements of F but only finitely many subfields of E including F, it follows that there are distinct elements a_1, a_2 of F such that $F(\alpha_0 + a_1\beta) = F(\alpha_0 + a_2\beta)$. Then $F(\alpha_0 + a_1\beta)$ contains $(\alpha_0 + a_1\beta) - (\alpha_0 + a_2\beta) = (a_1 - a_2)\beta$; hence it contains β, and hence also $\alpha_0 = (\alpha_0 + a_1\beta) - a_1\beta$; hence $F(\alpha_0 + \alpha_1\beta)$ includes $F(\alpha_0, \beta)$. But, since it is included in $F(\alpha_0, \beta)$, it follows that $F(\alpha_0 + a_1\beta) = F(\alpha_0, \beta)$. Hence $(F(\alpha_0 + a_1\beta): F) > (F(\alpha_0): F)$, which contradicts our choice of α_0.

Hence $E = F(\alpha_0)$ as asserted.

Let now H_1 and H_2 be subgroups of a group G. The **intersection** of H_1 and H_2, which we denote by $H_1 \cap H_2$, is the subset of G consisting of all the elements common to H_1 and H_2; it is easily seen to be a subgroup of G, and is in fact the largest subgroup of G included in both

H_1 and H_2. The subset of G consisting of all finite products of elements chosen from H_1 and H_2 is also a subgroup of G; it is the smallest subgroup of G including both H_1 and H_2 and is called the **subgroup of G generated by H_1 and H_2**.

Next suppose E_1 and E_2 are subfields of a field K. Their intersection, $E_1 \cap E_2$, is a subfield of K; it is the largest subfield of K included in both E_1 and E_2. The subfield $E_1(E_2)$ of K generated over E_1 by E_2 clearly coincides with the subfield $E_2(E_1)$ generated over E_2 by E_1. We call this subfield the **compositum** of E_1 and E_2; it is the smallest subfield of K including both E_1 and E_2.

Suppose now that E_1 and E_2 both include the field F and that K is a separable normal extension of finite degree over F with Galois group G. Then the compositum $E_1(E_2)$ and the intersection $E_1 \cap E_2$ of E_1 and E_2 are subfields of K including F; it is natural to enquire how the subgroups $\Gamma(E_1(E_2))$ and $\Gamma(E_1 \cap E_2)$ of G, which correspond to $E_1(E_2)$ and $E_1 \cap E_2$ in the Galois correspondence of Theorem 16.1, are related to the subgroups $\Gamma(E_1)$ and $\Gamma(E_2)$ corresponding to E_1 and E_2. Our next theorem answers this question.

THEOREM 18.2. *Let K be a separable normal extension of finite degree over a subfield F, with Galois group G. Let E_1 and E_2 be subfields of K including F. If $\Gamma(E_1) = H_1$ and $\Gamma(E_2) = H_2$ then $\Gamma(E_1(E_2)) = H_1 \cap H_2$ and $\Gamma(E_1 \cap E_2)$ is the subgroup of G generated by H_1 and H_2.*

Proof. (1) Since $H_0 = \Gamma(E_1(E_2))$ consists of those automorphisms of K which leave $E_1(E_2)$ fixed, it follows that if γ is any element of H_0 then γ acts like the identity on E_1 and also on E_2; thus γ belongs to H_1 and to H_2. So H_0 is included in $H_1 \cap H_2$. But conversely it follows at once from Theorem 8.1 that any automorphism which leaves both E_1 and E_2 fixed also leaves $E_1(E_2)$ fixed and hence belongs to H_0. Thus $H_1 \cap H_2$ is included in H_0.

Hence $H_0 = H_1 \cap H_2$ as required.

(2) Let H be the subgroup of G generated by H_1 and H_2. Since H consists of finite products of elements of H_1 and H_2, it is clear that every element of H acts like the identity on $E_1 \cap E_2$. That is to say, $\Phi(H)$ includes $E_1 \cap E_2$. Now let x be any element of K which does not belong to $E_1 \cap E_2$; say x does not belong to E_1. Then, since E_1 is the precise fixed field under H_1, there is at least one element γ_1 of H_1 such that $\gamma_1(x) \neq x$ and hence x is not left fixed by H. It follows that $\Phi(H) = E_1 \cap E_2$ and hence that $H = \Gamma(E_1 \cap E_2)$.

We now consider the situation in which K and E are subfields of a field L such that K is a separable normal extension of finite degree over a subfield F which is also a subfield of E and E is an arbitrary extension of F (possibly neither normal nor separable nor even finite). The next theorem, which deals with this situation, is traditionally known as Lagrange's *Theorem on Natural Irrationalities.*

THEOREM 18.3. *Let K and E be subfields of a field L. If K is a separable normal extension of finite degree over a subfield F which is also included in E, then $E(K)$ is a separable normal extension of finite degree over E. Further, the Galois group of $E(K)$ over E is isomorphic to the Galois group of K over the intersection of E and K.*

Proof. The fields involved in this discussion are illustrated in fig. 7.

Since K is a normal extension of F, it follows from Theorem 15.1 that K is a splitting field over F for a polynomial f in $P(F)$. Since K is a separable extension of F, the construction given in Theorem 15.1 actually leads to a separable polynomial f. Thus $K = F(\alpha_1, \ldots, \alpha_r)$, where $\alpha_1, \ldots, \alpha_r$ are the roots of f, and hence $E(K) = E(\alpha_1, \ldots, \alpha_r)$. It follows that $E(K)$ is a splitting field for f over E. Hence, by Theorem 15.1 and the Corollary to Theorem 13.7, $E(K)$ is a separable normal extension of finite degree over E.

Let now H be the Galois group of $E(K)$ over E. For each E-automorphism γ in H we have a monomorphism γ' of K into L, defined by setting $\gamma'(x) = \gamma(x)$ for all elements x of K; since γ acts like the identity on E, γ' acts like the identity on F. Further, since γ maps each root α_i of f onto some other root α_j of f, so also does γ'; hence γ' is a monomorphism of K into itself. But since γ is an automorphism of $E(K)$, every root α_i of f is the image of some root of f under γ and hence under γ'. This discussion shows that γ' is an F-automorphism of K. Thus we have a mapping ϕ of H into the Galois group G of K over F defined by setting $\phi(\gamma) = \gamma'$ for every element γ of H. It is easy to verify that ϕ is a group monomorphism of H into G; thus H is isomorphic to the subgroup $\phi(H)$ of G.

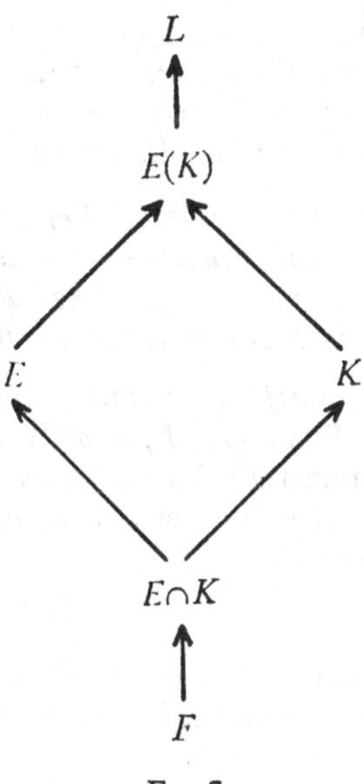

FIG. 7

To complete the proof we have to show that the fixed field under $\phi(H)$ is the subfield $E \cap K$ of K. It is clear, of course, that for every element γ of H the automorphism $\phi(\gamma)$ acts like the identity on $E \cap K$; so the fixed field under $\phi(H)$ includes $E \cap K$. Suppose now that x is an element of K which does not belong to $E \cap K$; then x does not belong to E. Since E is the precise fixed field under H (Theorem 15.5), there is an element γ_1 of H such that $\gamma_1(x) \neq x$. Thus $(\phi(\gamma_1))(x) \neq x$ and so x does not belong to the fixed field under $\phi(H)$. It follows that $E \cap K$ is the precise fixed field under $\phi(H)$, as asserted.

§ **19. Normal bases.** Let K be a separable normal extension of finite degree n over a subfield F; let $G = \{\tau_1, \ldots, \tau_n\}$ be the Galois group of K over F. A **normal basis** for K over F is a basis for K over F consisting of the images $\tau_1(x), \tau_2(x), \ldots, \tau_n(x)$ of a single element x of K under the elements of the Galois group. We shall deduce from our next theorem a criterion that the images of an element x should form a normal basis for K over F.

THEOREM 19.1. *Let K be a separable normal extension of finite degree n over a subfield F, with Galois group $G = \{\tau_1, \ldots, \tau_n\}$. The subset $\{x_1, \ldots, x_n\}$ of K is a basis for K over F if and only if the matrix $[\tau_i(x_j)]$ is non-singular.*

Proof. (1) Suppose the matrix $[\tau_i(x_j)]$ is non-singular.

Since $(K: F) = n$, if we can show that $\{x_1, \ldots, x_n\}$ is linearly independent over F, it will follow from Theorem 4.3 that this set is a basis for K over F. So suppose we have

$$a_1x_1 + \ldots + a_nx_n = 0,$$

where $a_1, \ldots, a_n$ are elements of F. Applying the F-automorphisms $\tau_1, \ldots, \tau_n$, we obtain

$$\tau_i(x_1)a_1 + \ldots + \tau_i(x_n)a_n = 0 \quad (i = 1, \ldots, n).$$

Since the matrix of coefficients $[\tau_i(x_j)]$ is non-singular, it follows from the theory of homogeneous linear equations that $a_1 = \ldots = a_n = 0$.

Thus $\{x_1, \ldots, x_n\}$ is linearly independent and so forms a basis, as required.

(2) Conversely, suppose that the matrix $[\tau_i(x_j)]$ is singular.

It follows again from the theory of homogeneous

linear equations that there exist elements $\alpha_1, \ldots, \alpha_n$ of K, not all zero, such that

$$\alpha_1\tau_i(x_1)+\ldots+\alpha_n\tau_i(x_n) = 0 \quad (i = 1, \ldots, n). \qquad (19.1)$$

According to Theorem 17.1, there exists an element α of K such that $S_{K/F}(\alpha)$ is non-zero. If α_k is non-zero, we multiply the equations (19.1) by $\alpha\alpha_k^{-1}$ and obtain

$$\beta_1\tau_i(x_1)+\ldots+\beta_n\tau_i(x_n) = 0 \quad (i = 1, \ldots, n), \qquad (19.2)$$

where $\beta_j = \alpha\alpha_k^{-1}\alpha_j$ $(j = 1, \ldots, n)$ and hence, in particular, $\beta_k = \alpha$. Now apply the F-automorphisms $\tau_1^{-1}, \ldots, \tau_n^{-1}$ to the equations (19.2); we obtain

$$\tau_i^{-1}(\beta_1)x_1+\ldots+\tau_i^{-1}(\beta_n)x_n = 0 \quad (i = 1, \ldots, n). \quad (19.3)$$

Adding these equations, and remembering that, as τ_i runs through the group G, so does τ_i^{-1}, we deduce that

$$S_{K/F}(\beta_1)x_1+\ldots+S_{K/F}(\beta_n)x_n = 0. \qquad (19.4)$$

Hence, since the coefficients of (19.4) belong to F and $S_{K/F}(\beta_k) = S_{K/F}(\alpha)$ is non-zero, the set $\{x_1, \ldots, x_n\}$ is linearly dependent over F and so does not form a basis.

This completes the proof.

COROLLARY. *The images of an element x under the automorphisms in the Galois group form a normal basis if and only if the matrix $[(\tau_i\tau_j)(x)]$ is non-singular.*

We now contend that every separable normal extension of finite degree has a normal basis. As in the case of the Primitive Element Theorem (18.1), separate proofs are required for finite and infinite fields. We shall consider only infinite fields; in this case the result follows from the next two theorems.

THEOREM 19.2. *Let K be a field, F a subfield of K with infinitely many elements; let f be a non-zero polynomial in $P_n(K)$. Then there are infinitely many ordered n-tuples $\boldsymbol{a} = (a_1, \ldots, a_n)$ of elements of F such that $\sigma_{\boldsymbol{a}}(f)$ is non-zero.*

Proof. We proceed by induction on n.

Suppose $n = 1$ and let f be a polynomial of degree d in $P_1(K) = P(K)$. Then f can have at most d roots in F; so there are infinitely many elements a of F which are not roots of f, i.e., such that $\sigma_a(f) \neq 0$.

Next suppose we have established that if g is any polynomial in $P_k(K)$ there are infinitely many ordered k-tuples $\boldsymbol{a}' = (a_1, \ldots, a_k)$ of elements of F such that $\sigma_{\boldsymbol{a}'}(g)$ is non-zero. Let f be any non-zero polynomial in $P_{k+1}(K)$; since, by definition, $P_{k+1}(K) = P(P_k(K))$, we may express f in the form

$$f = g_0 + g_1 X_{k+1} + \ldots + g_r X_{k+1}^r,$$

where $g_0, \ldots, g_r$ are polynomials in $P_k(K)$. Since f is not the zero polynomial, at least one of the polynomials $g_0, \ldots, g_r$ must be non-zero—say g_i. According to the inductive hypothesis, there are infinitely many ordered k-tuples $\boldsymbol{a}' = (a_1, \ldots, a_k)$ of elements of F such that $\sigma_{\boldsymbol{a}'}(g_i)$ is non-zero. For each of these k-tuples $\boldsymbol{a}'$, the polynomial

$$f_{\boldsymbol{a}'} = \sigma_{\boldsymbol{a}'}(g_0) + \sigma_{\boldsymbol{a}'}(g_1) X_{k+1} + \ldots + \sigma_{\boldsymbol{a}'}(g_r) X_{k+1}^r$$

is a non-zero polynomial in $P(K)$. Hence there are infinitely many elements a_{k+1} of F such that $\sigma_{a_{k+1}}(f_{\boldsymbol{a}'})$ is non-zero. But if we set $\boldsymbol{a} = (a_1, \ldots, a_k, a_{k+1})$ it is clear that

$$\sigma_{\boldsymbol{a}}(f) = \sigma_{a_{k+1}}(f_{\boldsymbol{a}'}).$$

Hence we see that there are infinitely many ordered $(k+1)$-tuples $\boldsymbol{a} = (a_1, \ldots, a_{k+1})$ of elements of F such that $\sigma_{\boldsymbol{a}}(f)$ is non-zero.

This completes the induction.

THEOREM 19.3. *Let K be a separable normal extension of finite degree n over a subfield F which contains infinitely many elements. Let $G = \{\tau_1, \ldots, \tau_n\}$ be the Galois group of K over F. If f is a polynomial in $P_n(K)$ such that, for every element α of K, the result of substituting $(\tau_1(\alpha), \ldots, \tau_n(\alpha))$ in f is zero, then f is the zero polynomial.*

Proof. Let $\{x_1, \ldots, x_n\}$ be a basis for K over F. Then, according to Theorem 19.1, the matrix $[\tau_i(x_j)]$ is non-singular, and hence has an inverse, $[t_{ij}]$ say.

Let g be the polynomial in $P_n(K)$ obtained from f when we replace each X_i by $\sum_{j=1}^{n} \tau_i(x_j)X_j \quad (i = 1, \ldots, n)$. If $\boldsymbol{a} = (a_1, \ldots, a_n)$ is any ordered n-tuple of elements of F, we set $\alpha = a_1x_1 + \ldots + a_nx_n$; then

$$\tau_i(\alpha) = a_1\tau_i(x_1) + \ldots + a_n\tau_i(x_n) \ (i = 1, \ldots, n).$$

We notice then that the result of substituting $\boldsymbol{a}$ in g is the same as the result of substituting $(\tau_1(\alpha), \ldots, \tau_n(\alpha))$ in f, and this, by hypothesis, is zero. It follows from Theorem 19.2 that g is the zero polynomial, since otherwise there would be infinitely many n-tuples $\boldsymbol{a}$ for which $\sigma_{\boldsymbol{a}}(g)$ is non-zero.

Let now $\boldsymbol{b} = (b_1, \ldots, b_n)$ be any ordered n-tuple of elements of F and set $c_j = \sum_{r=1}^{n} t_{jr}b_r \quad (j = 1, \ldots, n)$. Then we have

$$\sum_{j=1}^{n} \tau_i(x_j)c_j = \sum_{j=1}^{n} \sum_{r=1}^{n} \tau_i(x_j)t_{jr}b_r$$

$$= \sum_{r=1}^{n} \left(\sum_{j=1}^{n} \tau_i(x_j)t_{jr} \right) b_r = b_i,$$

since $\sum_{j=1}^{n} \tau_i(x_j)t_{jr}$ is the (i, r)th element of the unit matrix. Hence the result of substituting $\boldsymbol{c} = (c_1, \ldots, c_n)$ in g is the same as the result of substituting

$$\left(\sum_{j=1}^{n} \tau_1(x_j)c_j, \ldots, \sum_{j=1}^{n} \tau_n(x_j)c_j \right) = (b_1, \ldots, b_n)$$

in f. Thus $\sigma_{\boldsymbol{b}}(f) = \sigma_{\boldsymbol{c}}(g) = 0$.

It follows at once from Theorem 19.2 that f is the zero polynomial.

We can now prove the *Normal Basis Theorem* for the case of infinite fields.

THEOREM 19.4. *Let K be a separable normal extension of finite degree over a subfield F which contains infinitely many elements. Then there exists a normal basis for K over F.*

Proof. Let $G = \{\tau_1, \ldots, \tau_n\}$ be the Galois group of K over F. We write $\tau_i\tau_j = \tau_{p(i,j)}$. Since $\tau_i\tau_j = \tau_i\tau_k$ if and only if $j = k$, it follows that $p(i, j) = p(i, k)$ if and only if $j = k$; similarly, $p(h, j) = p(i, j)$ if and only if $h = i$. Consider now the polynomial f in $P_n(K)$ obtained by taking the determinant of the matrix whose (i, j)th element is $X_{p(i,j)}$. It follows from our remarks about $p(i, j)$ that X_1 occurs exactly once in each row and exactly once in each column of this matrix. If now we substitute the ordered n-tuple $(e, 0, \ldots, 0)$ in f we obtain the determinant of a matrix in which the identity element e of F occurs exactly once in each row and exactly once in each column; the determinant of such a matrix is either e or $-e$. Hence f is not the zero polynomial.

It now follows from Theorem 19.3 that there is at least one element x of K such that the result of substituting $(\tau_1(x), \ldots, \tau_n(x))$ in f is non-zero. But the result of this substitution is the determinant of the matrix $[\tau_i(\tau_j(x))]$; so this matrix is non-singular. Hence, by the Corollary to Theorem 19.1, $\{\tau_1(x), \ldots, \tau_n(x)\}$ is a normal basis for K over F.

Examples III

1. Let K be an algebraic extension of a subfield F. If for every element α of K the minimum polynomial of α over F has degree either 1 or 2, prove that K is a normal extension of F.

2. Let L be an algebraic extension of a subfield F, K a

subfield of L including F which is a separable normal extension of F. Prove that for each element α of L the minimum polynomial of α over F splits in $P(K)$ as a product of irreducible factors all of the same degree.

3. Let K be a separable normal extension of finite degree over a subfield F. If α is an element of K which has exactly r distinct images under the automorphisms in the Galois group of K over F, show that $(F(\alpha): F) = r$.

4. Let K be a splitting field over F for a separable polynomial f in $P(F)$; let G be the Galois group of K over F. Denote the set of roots of f in K by W, and for each F-automorphism τ in G define a mapping τ' of W into K by setting $\tau'(\alpha) = \tau(\alpha)$ for every root α of f in K. Show that τ' is a permutation of W (i.e., a one-to-one mapping of W onto itself). Thus we may consider G as a group of permutations of W. Prove that this group is transitive if and only if f is irreducible in $P(F)$. (A group G of permutations of a set W is said to be transitive if for every pair of elements α, β of W there exists an element τ in G such that $\tau(\alpha) = \beta$.)

5. Let K be a separable normal extension of finite degree over a subfield F; let G be the Galois group. If E is a subfield of K including F, corresponding to the subgroup H of G, show that two elements of G have the same effect on E if and only if they lie in the same left coset of G relative to H. Prove also that the number of F-automorphisms of E is equal to the index of H in its normaliser.

6. Construct subfields of $\mathbf{C}$ which are splitting fields over $\mathbf{Q}$ for the polynomials $X^4 - X^2 - 6$ and $X^3 - 3$; determine the Galois groups and the subfields of those splitting fields.

7. Let E be a separable extension of finite degree m over a subfield F. For each element α of E we define

elements $n_{E/F}(\alpha)$ and $s_{E/F}(\alpha)$ of F, which we call the **reduced norm** and **reduced trace** of α, as follows. Let the minimum polynomial of α over F be $X^k+a_1X^{k-1}+\ldots+a_k$; then set $n_{E/F}(\alpha)=(-1)^k a_k$ and $s_{E/F}(\alpha)=-a_1$. Prove that if $m = kl$, then $N_{E/F}(\alpha) = (n_{E/F}(\alpha))^l$ and $S_{E/F}(\alpha) = ls_{E/F}(\alpha)$.

8. Show that $\sqrt{2}+\sqrt{3}$ is a primitive element for $\mathbf{Q}(\sqrt{2}, \sqrt{3})$ over $\mathbf{Q}$. Find a primitive element for $\mathbf{Q}(\alpha, i)$ over $\mathbf{Q}$, where α is the positive real number such that $\alpha^4 = 2$.

9. Let α be the positive real number such that $\alpha^4 = 2$. Set $K = \mathbf{Q}(\alpha, i)$, $E = \mathbf{R}$. Determine the subfields $E(K)$ and $E \cap K$ of $\mathbf{C}$ and verify Lagrange's Theorem.

10. Show that $\{\omega, \omega^2\}$ is a normal basis for $\mathbf{Q}(\omega)$ over $\mathbf{Q}$, where $\omega = \frac{1}{2}(-1+i\sqrt{3})$. Find a normal basis for $\mathbf{Q}(i)$ over $\mathbf{Q}$.

CHAPTER IV

APPLICATIONS

§ **20. Finite fields.** We saw in § 1 that for every prime number p there exists a field with p elements, namely the field $\mathbf{Z}_p$ of residue classes of integers modulo p. Furthermore we deduce easily from Theorem 3.2 that every field consisting of p elements is isomorphic to $\mathbf{Z}_p$. If we follow our usual practice of regarding isomorphic fields as " essentially the same ", we may say that for each prime number p there exists one and " essentially only one " field with p elements. It is natural to wonder whether there are any other fields which consist of only finitely many elements and, if so, what can be said about their structure. We begin this investigation by proving a result which holds for all fields, finite or infinite.

THEOREM 20.1. *Let F be a field, M a finite subgroup of the multiplicative group F^* of non-zero elements of F. Then M is a cyclic group.*

Proof. Since multiplication in a field is commutative, M is a finite abelian group. Hence, according to the theory of such groups, M is isomorphic to a direct product $M_1 \times \ldots \times M_r$ of finite cyclic groups of orders $m_1, \ldots, m_r$ respectively, such that m_i is a factor of m_{i-1} $(i = 2, \ldots, r)$. It follows that the order of each element of M is a factor of m_1; so every element a of M is a root of the polynomial

$X^{m_1}-e$ (where e is the identity element of F and hence also of M). But this polynomial cannot have more than m_1 roots in F. Hence $M = M_1$, which is cyclic of order m_1.

We now establish a number of simple but important properties of finite fields.

THEOREM 20.2. *Let F be a finite field. Then* (1) *the characteristic of F is a prime number p and the number of elements q in F is a power of the characteristic*; (2) *the non-zero elements of F are all powers of a single one of them*; (3) *every element of F is a root of the polynomial X^q-X, and F is a splitting field of this polynomial over its prime field.*

Proof. (1) We have already proved, in the Corollary to Theorem 3.2, that the characteristic of F is a prime number p. Thus, according to Theorem 3.2 itself, the prime field F_0 of F is isomorphic to $\mathbf{Z}_p$. It is clear that F must be of finite degree over F_0, say $(F\colon F_0) = n$; let $\{x_1, \ldots, x_n\}$ be a basis for F over F_0. Then every element x of F can be expressed uniquely in the form

$$x = a_1x_1+\ldots+a_nx_n$$

where $a_1, \ldots, a_n$ are elements of F_0. Since there are p possible choices for each of the coefficients $a_i (i = 1, \ldots, n)$ it follows that there are precisely p^n elements of F, i.e., $q = p^n$.

(2) Since the multiplicative group F^* of non-zero elements of F is finite, it follows from Theorem 20.1 that F^* is cyclic; that is to say, all its elements are powers of a single one of them.

(3) Since F^* has $q-1$ elements, it follows that for every non-zero element x of F we have $x^{q-1} = e$ and hence $x^q = x$. But since $0^q = 0$ also, it follows that every element of F is a root of X^q-X. It follows that X^q-X

has q distinct roots in F and hence, since it has degree q, splits completely in $P(F)$. Clearly it cannot split completely in any subfield of F; so F is a splitting field for $X^q - X$ over F_0.

COROLLARY. *Any two finite fields having the same number of elements are isomorphic.*

Proof. Suppose F_1 and F_2 are fields with $q = p^n$ elements. Then their prime fields are isomorphic (since both are isomorphic to $\mathbf{Z}_p$). The result now follows from Theorem 11.2 since both F_1 and F_2 are splitting fields of $X^q - X$ over their prime fields.

Theorem 20.2 shows that we cannot construct finite fields with q_1 elements unless q_1 is a power of a prime number. But if q_1 is a power of a prime we shall deduce from the next theorem that there do exist fields with q_1 elements; by the corollary we have just proved, all such fields are isomorphic.

THEOREM 20.3. *Let F be a finite field with q elements, n a positive integer. Then there exist extensions of F of degree n, and all such extensions are isomorphic.*

Proof. As in the first part of Theorem 20.2 we can show that every extension of F of degree n contains $q_1 = q^n$ elements; so, by the Corollary, all such extensions are isomorphic.

To show that there exists at least one extension of degree n we construct a splitting field K over F for the polynomial $f = X^{q_1} - X$; we shall consider F as a subfield of K. Now the derivative of f is $Df = q_1 X^{q_1 - 1} - e = -e$, since q_1 is divisible by the characteristic of F; hence f and Df have no non-constant common factor (because Df has no non-constant factor at all). It follows from Theorem 13.1 that f has exactly $q_1 = q^n$ distinct roots in K.

Let E be the subset of K consisting of these q^n distinct roots. Then E includes F; for if a is any element of F it follows from Theorem 20.2 that $a^q = a$ and hence, by iteration, $a^{q_1} = a$. Next we show that E is actually a field. For, if x and y are roots of f in K, we may apply the Corollary of Theorem 3.3 and obtain

$$(x+y)^{q_1} = x^{q_1} + y^{q_1} = x+y,$$

$$(xy)^{q_1} = x^{q_1} y^{q_1} = xy,$$

$$(-x)^{q_1} = -x^{q_1} = -x,$$

and, if x is non-zero,

$$(x^{-1})^{q_1} = (x^{q_1})^{-1} = x^{-1};$$

thus $x+y$, xy, $-x$ and (when x is non-zero) x^{-1} lie in E.

Hence E is an extension of F with q^n elements.

COROLLARY 1. *If p is a prime number and n is any positive integer, there exists one (and essentially only one) field with p^n elements.*

Proof. We have only to take $F = \mathbf{Z}_p$ in the theorem.

Finite fields are sometimes called **Galois fields**, and the (essentially unique) field with p^n elements is often denoted by $\mathbf{GF}(p^n)$.

COROLLARY 2. *If E is an extension of finite degree over a finite field F, then there exists a primitive element for E over F.*

Proof. Let α be a generator of the cyclic multiplicative group of non-zero elements of E. Then clearly $E = F(\alpha)$.

We remark that Corollary 2 completes the proof of the Primitive Element Theorem (18.1).

COROLLARY 3. *If F is any finite field, the polynomial ring $P(F)$ contains irreducible polynomials of every degree.*

Proof. Let n be any positive integer. Let E be an extension of F of degree n; we have just seen in Corollary 2 that there is an element α of E such that $E = F(\alpha)$. Let m be the minimum polynomial of α over F, Then m is an irreducible polynomial in $P(F)$ and, according to Theorem 8.2, $\partial m = (F(\alpha): F) = n$.

Let F be a finite field with q elements, K an extension of degree n over F: we shall suppose that K includes F. Since all finite fields are perfect (Corollary to Theorem 13.3), K is a separable extension of F; since K is a splitting field over F of a polynomial in $P(F)$, K is a normal extension of F (Theorem 15.1). We now determine the Galois group of K over F.

THEOREM 20.4. *Let F be a finite field with q elements, K an extension of degree n over F including F. Then the Galois group of K over F is cyclic of order n and is generated by the F-automorphism τ defined by setting $\tau(x) = x^q$ for all elements x of K.*

Proof. Since $(K: F) = n$, the Galois group G of K over F is of order n.

We now show that the mapping τ is actually an F-automorphism of K. Certainly τ is a homomorphism of K into itself (Corollary to Theorem 3.3). Since $\tau(e) = e \neq 0$, τ is not the zero homomorphism, and hence (Theorem 3.1) is a monomorphism. But, because K is a finite field, it must therefore be an automorphism. Finally, for every element a of F it follows from Theorem 20.2 that

$$\tau(a) = a^q = a;$$

thus τ is an F-automorphism of K.

Let d be the order of τ in G; of course $d \leqq n$. Then, for every element x of K, we have $x = \varepsilon(x) = \tau^d(x) = x^{q^d}$. So all the q^n elements of K are roots of the polynomial $X^{q^d} - X$; hence $q^n \leqq q^d$, i.e., $n \leqq d$. It follows that $d = n$; thus τ has n distinct powers in G. So all the elements of G are powers of τ, i.e., G is a cyclic group generated by τ.

§ **21. Cyclotomic extensions.** Let F be a field with identity element e; for every positive integer m we denote by k_m the polynomial $X^m - e$ in $P(F)$. An extension of F is called a **cyclotomic extension** if it is a splitting field over F of one of the polynomials k_m.

Suppose now that F has non-zero characteristic p; then every positive integer m can be expressed in the form $m = p^r m_1$, where $r \geqq 0$ and p does not divide m_1. Then we have $k_m = X^m - e = (X^{m_1} - e)^{p^r}$; and so every splitting field of k_{m_1} over F is also a splitting field for k_m over F. Hence in the case where F has non-zero characteristic we may restrict our attention to those polynomials k_m for which m is not divisible by the characteristic. For such a polynomial, $Dk_m = mX^{m-1}$ is not the zero polynomial; the only non-constant monic factors of Dk_m are powers of X, none of which is a factor of k_m. It follows from Theorem 13.1 that k_m is a separable polynomial. Hence all cyclotomic extensions of F are separable and normal.

Let K_m be a splitting field for k_m over F, where m is not divisible by the characteristic of F if this is non-zero; we shall suppose that K_m includes F as a subfield. As we have just seen, the m roots of k_m in K_m are all distinct; we call them the mth **roots of unity** in K_m and denote them by $\zeta_1, \ldots, \zeta_m$. These mth roots of unity clearly form a subgroup of the multiplicative group of non-zero elements of K_m; for, if ζ_i and ζ_j are mth roots of unity in K_m, we have $(\zeta_i\zeta_j)^m = \zeta_i^m\zeta_j^m = e$ and $(\zeta_i^{-1})^m = e$ and so $\zeta_i\zeta_j$ and ζ_i^{-1} are also mth roots of unity. According to Theorem 20.1, the group of mth roots of unity in K_m is a cyclic group. Any one of the roots of unity which can be taken as a generator of this group is called a **primitive** mth root of unity in K_m. If ζ is a primitive mth root of unity, then it is easily verified that the order of ζ^r in the group of mth roots of unity is m/d, where d is the highest common factor of r and m; it follows that the primitive mth roots of unity are precisely those of the form ζ^r where r is relatively prime to m. In particular, if m is a prime number, every mth root of unity except the identity element e is a primitive mth root of unity. It is clear that any primitive mth root of unity ζ may be taken as a primitive element for K_m over F: that is to say, $K_m = F(\zeta)$.

Since cyclotomic extensions are separable and normal it is natural that we should try to get some information about the Galois groups of such extensions. For this purpose we recall to mind the group $\mathbf{R}_m$ which we introduced in Example 4 of § 1. The elements of this group $\mathbf{R}_m$ are the residue classes modulo m consisting of integers which are relatively prime to m, and we form the product of two such relatively prime residue classes C_1, C_2 by choosing integers n_1, n_2 from C_1, C_2 respectively and defining C_1C_2 to be the residue class containing n_1n_2. We denote the order of $\mathbf{R}_m$ by $\phi(m)$, and it can be shown that

$$\phi(m) = m\prod\left(1-\frac{1}{p}\right),$$

where the product is extended over the prime numbers p which divide m.

We can now state our first theorem about the Galois group of a cyclotomic extension.

THEOREM 21.1. *Let F be a field, m a positive integer which is not divisible by the characteristic of F if this is non-zero. Let K_m be a splitting field for k_m over F including F. Then the Galois group G of K_m over F is isomorphic to a subgroup of $\mathbf{R}_m$.*

Proof. Let ζ be a primitive mth root of unity in K_m. If τ is any element of G, $\tau(\zeta)$ is also a primitive mth root of unity: hence there is an integer n_τ relatively prime to m such that $\tau(\zeta) = \zeta^{n_\tau}$. We now define a mapping θ of

G into $\mathbf{R}_m$ by setting $\theta(\tau) =$ the residue class of n_τ modulo m. Then θ is a homomorphism; for if τ and ρ are elements of G we have

$$\zeta^{n_{\tau\rho}} = (\tau\rho)(\zeta) = \tau(\rho(\zeta)) = \tau(\zeta^{n_\rho}) = (\tau(\zeta))^{n_\rho} = \zeta^{n_\tau n_\rho},$$

whence $n_{\tau\rho} \equiv n_\tau n_\rho$ mod. m, and hence $\theta(\tau\rho) = \theta(\tau)\theta(\rho)$. Next, θ is one-to-one; for if $\tau \neq \rho$ then $\tau(\zeta) \neq \rho(\zeta)$, i.e., $\zeta^{n_\tau} \neq \zeta^{n_\rho}$ and hence n_τ and n_ρ lie in different residue classes modulo m.

It follows that G is isomorphic to the subgroup $\theta(G)$ of $\mathbf{R}_m$.

The conclusion of this theorem clearly cannot be refined unless we are given some further information about the field F. For example, if F already contains a primitive mth root of unity, then $K_m = F$ and the Galois group G consists of the identity element alone. Again, if F is a finite field with q elements and $m = q^n - 1$, the results of § 20 show that K_m has degree n over F and the Galois group G is a cyclic subgroup of order n in $\mathbf{R}_m$. We shall show in the next section that if $F = \mathbf{Q}$, the field of rational numbers, then the Galois group of K_m over F is in fact isomorphic to $\mathbf{R}_m$ itself.

For the moment, however, let F be an arbitrary field and K_m a splitting field for k_m over F including F (as usual we assume that m is not divisible by the characteristic of F if this is non-zero). If d is any factor of m, $1 \leqq d \leqq m$, the polynomial $k_d = X^d - e$ is a factor of $k_m = X^m - e$ and hence included among the mth roots of unity in K_m there are d distinct dth roots of unity and in particular $\phi(d)$ primitive dth roots of unity. Thus for each factor d of m we may define the polynomial Φ_d in $P(K_m)$ by setting

$$\Phi_d = \prod(X - \zeta_d),$$

where the product is taken over all the *primitive* dth roots of unity ζ_d in K_m; then $\partial\Phi_d = \phi(d)$. Since every mth root of unity ζ is a primitive dth root of unity for some factor d of m (d is the order of ζ in the multiplicative group of mth roots of unity), it follows that

$$k_m = X^m - e = \prod \Phi_d,$$

where the product is extended over all the factors d of m, $1 \leqq d \leqq m$. The polynomial Φ_m is called the mth **cyclotomic polynomial.**

THEOREM 21.2. *For every positive integer* m, *the coefficients of the* mth *cyclotomic polynomial belong to the prime field of* F; *if* F *has characteristic zero and we identify the prime field with* $\mathbf{Q}$, *these coefficients are integers.*

Proof. We proceed by induction on m.

Certainly $\Phi_1 = X - e$ has coefficients in the prime field. If the prime field is $\mathbf{Q}$ we have $e = 1$ and so $\Phi_1 = X - 1$ has integer coefficients.

Suppose now that the result of the theorem holds for all factors d of m such that $d < m$. Then we have

$$X^m - e = \Phi_m \prod{}' \Phi_d,$$

where the product Π' on the right-hand side is extended over all the factors d of m such that $1 \leqq d < m$. By hypothesis, all the factors in this product have coefficients in the prime field; $X^m - e$ has coefficients in the prime field. Hence so does Φ_m. In the case where the prime field is $\mathbf{Q}$, every factor in the product has integer coefficients and leading coefficient 1; when we divide a polynomial with integer coefficients by a polynomial with integer coefficients and leading coefficient 1 the quotient has integer coefficients. Thus Φ_m has integer coefficients.

Example. We shall compute Φ_{18}.

Since the factors of 18 are 1, 2, 3, 6, 9 and 18, we have

$$X^{18} - 1 = \Phi_1 \Phi_2 \Phi_3 \Phi_6 \Phi_9 \Phi_{18}.$$

Similarly, since the factors of 9 are 1, 3 and 9, we have

$$X^9 - 1 = \Phi_1\Phi_3\Phi_9.$$

Hence

$$X^9 + 1 = \Phi_2\Phi_6\Phi_{18}.$$

The occurrence of Φ_6 in this product suggests that we examine $X^6 - 1$; we have

$$X^6 - 1 = \Phi_1\Phi_2\Phi_3\Phi_6.$$

Now clearly

$$X^3 - 1 = \Phi_1\Phi_3.$$

So we have

$$\Phi_2\Phi_6 = X^3 + 1.$$

Thus finally

$$\Phi_{18} = (X^9+1)/(X^3+1) = X^6 - X^3 + 1.$$

§ 22. Cyclotomic extensions of the rational number field. We now approach the task of proving the result mentioned above (just after Theorem 21.1) that if $F = \mathbf{Q}$ then the Galois group of K_m over F is isomorphic to the multiplicative group $\mathbf{R}_m$ of residue classes modulo m relatively prime to m. For this purpose we require some preliminary results concerning polynomials with rational coefficients and polynomials with integer coefficients. Let f be a polynomial in $P(\mathbf{Z})$; then the **content** of f is the positive highest common factor of the coefficients of f; if the content of f is 1 we say that f is a **primitive** polynomial. Clearly if f is any polynomial in $P(\mathbf{Z})$ we may write $f = cf_1$, where c is the content of f and f_1 is a primitive polynomial in $P(\mathbf{Z})$.

THEOREM 22.1. *If a polynomial with integer coefficients can be expressed as a product of two polynomials with rational coefficients, then it can be expressed as a product of two polynomials with integer coefficients.*

Proof. Let f be a polynomial in $P(\mathbf{Z})$ and g_1, g_2 two polynomials in $P(\mathbf{Q})$ such that $f = g_1 g_2$. Let m_1, m_2 be the least common multiples of the denominators of the coefficients of g_1, g_2 respectively. Then $h_1 = m_1 g_1$ and $h_2 = m_2 g_2$ are polynomials in $P(\mathbf{Z})$. Let c_1 and c_2 be the contents of h_1 and h_2 and write $h_1 = c_1 h'_1$, $h_2 = c_2 h'_2$, where h'_1 and h'_2 are primitive polynomials in $P(\mathbf{Z})$. Then we have

$$(m_1 m_2)f = (c_1 c_2)(h'_1 h'_2).$$

We claim that $h'_1 h'_2$ is a primitive polynomial. So let p be any prime number. Since $h'_1 = a_0 + a_1 X + a_2 X^2 + \ldots$ and $h'_2 = b_0 + b_1 X + b_2 X^2 + \ldots$ are primitive polynomials, each has at least one coefficient which is not divisible by p; let a_i and b_j be the first coefficients of h'_1 and h'_2 respectively which are not divisible by p. Then the coefficient of X^{i+j} in $h'_1 h'_2$ is $\Sigma a_\nu b_\mu$, where the summation extends over all pairs of indices (ν, μ) such that $\nu + \mu = i + j$. If $\nu \neq i$, $\mu \neq j$ and $\nu + \mu = i + j$, then either $\nu < i$ or $\mu < j$ and hence either a_ν is divisible by p or b_μ is divisible by p; thus all the terms in the summation are divisible by p except for $a_i b_j$ itself, and so the sum is not divisible by p. It follows that for every prime number p, $h'_1 h'_2$ has at least one coefficient which is not divisible by p; hence the positive highest common factor of the coefficients of $h'_1 h'_2$ is 1, i.e., $h'_1 h'_2$ is primitive.

It follows that $c_1 c_2$ is the content of $(m_1 m_2)f$. But $m_1 m_2$ is clearly a factor of the content of $(m_1 m_2)f$; hence $c_1 c_2 / m_1 m_2$ is an integer k, say. Then $f = (k h'_1) h'_2$ is a factorisation of f in $P(\mathbf{Z})$.

COROLLARY. *If f is a monic factor of $X^m - 1$ in $P(\mathbf{Q})$, then f has integer coefficients.*

We now consider pairs of polynomials

$$f = a_0 + a_1 X + a_2 X^2 + \ldots, \quad g = b_0 + b_1 X + b_2 X^2 + \ldots$$

in $P(\mathbf{Z})$; we say that f is congruent to g modulo a positive integer n, and we write $f \equiv g$ mod. n, if corresponding coefficients of f and g are congruent modulo n, i.e., if $a_i \equiv b_i$ mod. n $(i = 0, 1, 2, \ldots)$.

If k is any positive integer, we shall denote by $f^{(k)}$ the polynomial obtained from f when we replace X by X^k, i.e.,

$$f^{(k)} = a_0 + a_1X^k + a_2X^{2k} + \ldots.$$

It is easy to verify that the mapping ρ_k of $P(\mathbf{Z})$ into itself defined by setting $\rho_k(f) = f^{(k)}$ for every polynomial f in $P(\mathbf{Z})$ is a monomorphism.

Let now f be any polynomial in $P(\mathbf{Z})$, p any prime number. Then we claim that $f^{(p)} \equiv f^p$ mod. p. To see this, let $f = a_0 + a_1X + \ldots + a_qX^q$. Then

$$f^p = a_0^p + a_1^pX^p + \ldots + a_q^pX^{qp} + s,$$

where s is a polynomial in $P(\mathbf{Z})$ each of whose coefficients has a factor of the form $p!/r_0!r_1!\ldots r_q!$ $(0 \leqq r_0, r_1, \ldots, r_q < p, r_0 + r_1 + \ldots + r_q = p)$ and so is divisible by p. Hence

$$f^p \equiv a_0^p + a_1^pX^p + \ldots + a_q^pX^{qp} \text{ mod. } p.$$

Let now a be any integer, A its residue class modulo p. Then A is an element of $\mathbf{Z}_p$, which is a finite field with p elements. So it follows from Theorem 20.2 that $A^p = A$, and hence, since A^p is the residue class of a^p modulo p, we deduce that $a^p \equiv a$ mod. p. Consequently

$$f^p \equiv a_0 + a_1X^p + \ldots + a_qX^{qp} = f^{(p)} \text{ mod. } p,$$

as asserted. We use this elementary result in the proof of the following theorem.

THEOREM 22.2. *Let f be any factor of $X^m - 1$ in $P(\mathbf{Z})$. If k is any positive integer relatively prime to m, then f is a factor of $f^{(k)}$ in $P(\mathbf{Z})$.*

Proof. Since f is a factor of $X^m - 1$ in $P(\mathbf{Z})$, its leading coefficient must be 1 or -1. Thus, if i is any positive integer and we write

$$f^{(i)} = fq_i + r_i, \quad \partial r_i < \partial f,$$

the polynomials q_i and r_i lie in $P(\mathbf{Z})$.

We now show that for every positive integer i we have $r_{m+i} = r_i$. So let $f = X^t + a_1 X^{t-1} + \ldots + a_t$; then we have

$$f^{(m+i)} - f^{(i)} = (X^{t(m+i)} - X^{ti}) + a_1(X^{(t-1)(m+i)} - X^{(t-1)i}) + \ldots$$
$$= X^{ti}(X^{tm} - 1) + a_1 X^{(t-1)i}(X^{(t-1)m} - 1) + \ldots$$

All the terms on the right-hand side are divisible by $X^m - 1$ and hence by f. It follows that

$$f^{(m+i)} - f^{(i)} = fq_{m+i} + r_{m+i} - fq_i - r_i$$

is divisible by f and hence $r_{m+1} - r_i$ is divisible by f; but $\partial(r_{m+i} - r_i) < \partial f$ and so it follows that $r_{m+i} - r_i$ is the zero polynomial, i.e., $r_{m+i} = r_i$, as asserted. By induction $r_{lm+i} = r_i$ for $l = 0, 1, 2, \ldots$.

It follows that if i is any positive integer then r_i is one of the m polynomials $r_1, \ldots, r_m$. Let π be the greatest prime number which is a factor of any non-zero coefficient of any of these polynomials, setting $\pi = 1$ if there are no non-zero coefficients or if all the non-zero coefficients are either 1 or -1.

Let now p be any prime number greater than π. We have just seen that $f^p \equiv f^{(p)}$ mod. p; so there is a polynomial g in $P(\mathbf{Z})$ such that $f^p = f^{(p)} + pg$. In the usual way we may write $g = fh + s$, where h and s are polynomials in $P(\mathbf{Z})$ and $\partial s < \partial f$. Thus we have

$$f^p = fq_p + r_p + p(fh + s),$$

from which we deduce that $r_p + ps$ is divisible by f; but, since the degree of this polynomial is less than the degree of f, we deduce that $r_p = -ps$. Since $p > \pi$, however, p is not a factor of any non-zero coefficient of r_p; so we have here a contradiction unless r_p and s are equal to the zero polynomial.

We have thus established that, if p is any prime number greater than π, f is a factor of $f^{(p)}$.

It now follows easily by induction that if $p_1, \ldots, p_n$ are n prime numbers greater than π (not necessarily distinct) then f is a factor of $f^{(p_1 \ldots p_n)}$. Namely, suppose we have established that f is a factor of $f^{(p_1 \ldots p_{n-1})}$; then $f^{(p_n)}$ is a factor of $(f^{(p_1 \ldots p_{n-1})})^{(p_n)} = f^{(p_1 \ldots p_n)}$ and since f is a factor of $f^{(p_n)}$ the desired result follows.

Now let k be any positive integer relatively prime to m. We denote by l the product of all the prime numbers $\leqq \pi$ which do not divide k (setting $l = 1$ if there are no such prime numbers). Then all the prime factors of $k+ml$ are greater than π. Hence f is a factor of $f^{(k+ml)}$, i.e., r_{k+ml} is the zero polynomial and so, finally, r_k is the zero polynomial.

Thus f is a factor of $f^{(k)}$, as asserted.

We now return to our problem of determining the Galois group of a splitting field of $k_m = X^m - 1$ over the field of rational numbers.

THEOREM 22.3. *Let K_m be a splitting field of k_m over* **Q** *which includes* **Q**. *Then the Galois group G of K_m over* **Q** *is isomorphic to the multiplicative group* $\mathbf{R}_m$ *of residue classes modulo m which are relatively prime to m.*

Proof. Let ζ be a primitive mth root of unity in K_m. In Theorem 21.1 we set up a monomorphism θ of G into $\mathbf{R}_m$ by remarking that for each automorphism τ in G we have $\tau(\zeta) = \zeta^{n_\tau}$ where n_τ is relatively prime to m, and then setting $\theta(\tau) =$ the residue class of n_τ modulo m. We shall now show that θ is an epimorphism.

Let f be the minimum polynomial of ζ over **Q**. Then f is a monic factor of $X^m - 1$ in $P(\mathbf{Q})$ and hence, according to the Corollary to Theorem 22.1, f has integer coefficients.

Now let C be any relatively prime residue class modulo m; let s be any integer in C. It follows from Theorem 22.2 that f is a factor of $f^{(s)}$—say $f^{(s)} = fq_s$. Applying the substitution homomorphism σ_ζ we see that

$$\sigma_\zeta(f^{(s)}) = \sigma_\zeta(f)\sigma_\zeta(q_s) = 0.$$

But the result of substituting ζ in $f^{(s)}$ is the same as the result of substituting ζ^s in f; so ζ^s is a root of f. Since f is an irreducible polynomial in $P(\mathbf{Q})$ and K_m is a normal extension of $\mathbf{Q}$, it follows from Corollary 3 to Theorem 15.1 that there is a $\mathbf{Q}$-automorphism τ of K_m which maps ζ onto ζ^s. Clearly $\theta(\tau) = C$, and hence θ is an epimorphism, as asserted.

This completes the proof.

COROLLARY. *The cyclotomic polynomials* Φ_m *are all irreducible in* $P(\mathbf{Q})$.

Proof. Let ζ be a primitive mth root of unity in a splitting field K_m of $X^m - 1$ over $\mathbf{Q}$; let f be the minimum polynomial of ζ over $\mathbf{Q}$. We have just seen that each of the $\phi(m)$ primitive roots of unity in K_m is a root of f; hence Φ_m is a factor of f. Since Φ_m is a monic polynomial, it follows that $\Phi_m = f$ and so is irreducible.

§ **23. Cyclic extensions.** Let F be a subfield of a field K. Then K is said to be a **cyclic extension** of F if it is a separable normal extension of finite degree over F with cyclic Galois group. In this section we shall obtain some of the simple classical results about such extensions. Our first theorem appeared in Hilbert's *Zahlbericht* (*Die Theorie der algebraischen Zahlkörper*, 1897) and is traditionally referred to as *Hilbert's Theorem* 90.

THEOREM 23.1. *Let* K *be a cyclic extension of a subfield* F; *let* τ *be a generator of the Galois group of* K *over* F. *If* x *is an element of* K, *then* $N_{K/F}(x) = e$ *(the identity of* F*)*

if and only if there is an element y in K such that $x = y/\tau(y)$, and $S_{K/F}(x) = 0$ if and only if there is an element z in K such that $x = z - \tau(z)$.

Proof. Suppose $(K: F) = n$; then $\tau^n = \varepsilon$, the identity automorphism.

(1) Suppose $x = y/\tau(y)$. Then

$$N_{K/F}(x) = \varepsilon(x)\tau(x)\tau^2(x)\dots\tau^{n-1}(x)$$

$$= \frac{y}{\tau(y)}\frac{\tau(y)}{\tau^2(y)}\frac{\tau^2(y)}{\tau^3(y)}\dots\frac{\tau^{n-1}(y)}{\tau^n(y)} = e.$$

Similarly, if $x = z - \tau(z)$, we have $S_{K/F}(x) = 0$.

(2) Conversely, suppose that

$$N_{K/F}(x) = \varepsilon(x)\tau(x)\tau^2(x)\dots\tau^{n-1}(x) = e.$$

Then x is clearly non-zero and it follows at once that $x^{-1} = \tau(x)\tau^2(x)\dots\tau^{n-1}(x)$.

Next, since the set of automorphisms $\{\varepsilon, \tau, \tau^2, \dots, \tau^{n-1}\}$ is linearly independent over K (Theorem 14.1), the mapping

$$\varepsilon + x\tau + x\tau(x)\tau^2 + \dots + x\tau(x)\dots\tau^{n-2}(x)\tau^{n-1}$$

is not the zero mapping of K into itself. That is to say, there is an element t of K such that

$$y = t + x\tau(t) + x\tau(x)\tau^2(t) + \dots + x\tau(x)\dots\tau^{n-2}(x)\tau^{n-1}(t)$$

is non-zero. Applying the automorphism τ we obtain

$$\tau(y) = \tau(t) + \tau(x)\tau^2(t) + \tau(x)\tau^2(x)\tau^3(t)$$
$$+ \dots + \tau(x)\tau^2(x)\dots\tau^{n-1}(x)t = x^{-1}y.$$

Thus $x = y/\tau(y)$.

Similarly, suppose

$$S_{K/F}(x) = x + \tau(x) + \tau^2(x) + \dots + \tau^{n-1}(x) = 0.$$

Then of course $\tau(x) + \tau^2(x) + \dots + \tau^{n-1}(x) = -x$.

We saw in Theorem 17.1 that $S_{K/F}$ is not the zero mapping; so let t be an element of K such that $S_{K/F}(t)$

is non-zero, and consider the element

$$z_1 = x\tau(t)+(x+\tau(x))\tau^2(t)+\dots$$
$$+(x+\tau(x)+\dots+\tau^{n-2}(x))\tau^{n-1}(t).$$

Applying the automorphism τ we obtain

$$\tau(z_1) = \tau(x)\tau^2(t)+(\tau(x)+\tau^2(x))\tau^3(t)+\dots$$
$$+(\tau(x)+\tau^2(x)+\dots+\tau^{n-1}(x))t$$
$$= \tau(x)\tau^2(t)+(\tau(x)+\tau^2(x))\tau^3(t)+\dots-xt.$$

Hence we have

$$z_1-\tau(z_1) = x(t+\tau(t)+\tau^2(t)+\dots+\tau^{n-1}(t)) = xS_{K/F}(t).$$

Since $S_{K/F}(t)$ lies in F and hence is left fixed by τ, it follows that if we write $z = z_1/S_{K/F}(t)$, then $x = z-\tau(z)$.

Before passing on, we remark that the expression of an element x of K in the form $x = y/\tau(y)$ is not unique; but clearly $y/\tau(y) = y_1/\tau(y_1)$ if and only if $y/y_1 = \tau(y/y_1)$, i.e., if and only if y/y_1 belongs to F. Similarly we have $z-\tau(z) = z_1-\tau(z_1)$ if and only if $z-z_1$ belongs to F.

We shall now study cyclic extensions of degree n over a field F under the hypothesis that the polynomial

$$k_n = X^n-e$$

splits completely in $P(F)$; this is equivalent to saying that F contains a primitive nth root of unity. In this situation we shall show that the cyclic extensions of degree n are splitting fields of irreducible binomials, i.e., polynomials of the form X^n-a.

THEOREM 23.2. *Let F be a field and let n be a positive integer not divisible by the characteristic of F (if this is non-zero). Suppose that k_n splits completely in $P(F)$. If K is a cyclic extension of degree n over F including F, there exists an element a of F such that X^n-a is irreducible in $P(F)$ and K is generated over F by a root of X^n-a.*

Proof. Let τ be a generator of the cyclic Galois group G of K over F.

Since k_n splits completely in $P(F)$, F contains a primitive nth root of unity, ζ say. Then ζ is left fixed by all n elements of G, since they act like the identity on F, and so $N_{K/F}(\zeta) = \zeta^n = e$. It follows from Theorem 23.1 that there is an element α of K such that $\zeta = \alpha/\tau(\alpha)$; hence $\tau(\alpha) = \zeta^{-1}\alpha$. It follows by induction that for $k = 0, 1, \ldots, n-1$ we have $\tau^k(\alpha) = \zeta^{-k}\alpha$. Thus the subgroup of G which leaves α (and hence $F(\alpha)$) fixed consists of the identity element ε alone. Hence, according to Theorem 16.1, $(K\colon F(\alpha)) = 1$ and so $K = F(\alpha)$.

Next we remark that $e = \zeta^n = (\alpha/\tau(\alpha))^n = \alpha^n/\tau(\alpha^n)$; so $\tau(\alpha^n) = \alpha^n$, and hence, by induction, $\tau^k(\alpha^n) = \alpha^n$ for $k = 0, 1, \ldots, n-1$. Thus α^n belongs to the fixed field under G, which is of course F itself. Set $\alpha^n = a$. Then α is a root of the polynomial $X^n - a$ in $P(F)$; hence the minimum polynomial m of α over F is a factor of $X^n - a$. But $\partial m = (F(\alpha)\colon F) = n$ and m is a monic polynomial; so $m = X^n - a$ and hence $X^n - a$ is irreducible in $P(F)$.

It is clear from the form of the roots $\tau^k(\alpha) = \zeta^{-k}\alpha$ of $X^n - a$ that they all belong to K; hence K is in fact a splitting field of $X^n - a$ over F.

We now show that, under the same assumption that k_n splits completely in $P(F)$, the splitting fields of all binomials $X^n - a$ are cyclic extensions of F (though not necessarily of degree n).

THEOREM 23.3. *Let F be a field and let n be a positive integer not divisible by the characteristic of F (if this is non-zero). Suppose that k_n splits completely in $P(F)$. If a is any non-zero element of F and K is a splitting field of $X^n - a$ over F including F, then K is a cyclic extension of F, generated over F by any one of the roots of $X^n - a$.*

Proof. Since $D(X^n - a) = nX^{n-1}$ is not the zero polynomial, $X^n - a$ has no non-constant factor in common with its derivative and hence is a separable polynomial. Thus K is a separable normal extension of F.

Let α and β be any two roots of $X^n - a$ in K. Then $(\alpha/\beta)^n = e$; so α/β is an nth root of unity in K and hence lies in F. Thus β lies in $F(\alpha)$; so K, which is generated over F by the set of all the roots of $X^n - a$, coincides with $F(\alpha)$.

Let G be the Galois group of K over F; since every element of K can be expressed as a polynomial in α with coefficients in F (Theorem 8.2), it follows that each F-automorphism τ of K is completely determined once we know its action on α. Since $\tau(\alpha)$ is also a root of $X^n - a$, then, as we saw in the preceding paragraph, there exists an nth root of unity ζ_τ in F such that $\tau(\alpha) = \zeta_\tau \alpha$.

This allows us to define a mapping ϕ of G into the group of nth roots of unity in F; namely, we set $\phi(\tau) = \zeta_\tau$ for all automorphisms τ in G. This mapping is clearly one-to-one; but it is also a homomorphism, for if τ and ρ are any two elements of G we have

$$\phi(\tau\rho) = \frac{(\tau\rho)(\alpha)}{\alpha} = \frac{\tau(\rho(\alpha))}{\alpha} = \frac{\tau(\alpha)}{\alpha}\,\frac{\tau(\rho(\alpha))}{\tau(\alpha)}$$

$$= \frac{\tau(\alpha)}{\alpha}\,\frac{\rho(\alpha)}{\alpha} = \phi(\tau)\phi(\rho)$$

(we use the fact that $\tau(\rho(\alpha)/\alpha) = \rho(\alpha)/\alpha$ since $\rho(\alpha)/\alpha$ belongs to F). Thus ϕ is an isomorphism of the Galois group G onto the subgroup $\phi(G)$ of the group of nth roots of unity in F; but this is a finite subgroup of the multiplicative group of F and hence is cyclic, by Theorem 20.1.

Thus G is a cyclic group as asserted.

Next let F be a field of non-zero characteristic p; we proceed to study cyclic extensions of degree p over F.

THEOREM 23.4. *Let K be a cyclic extension of degree p over a subfield F of non-zero characteristic p. Then there* *exists an element a of F such that $X^p - X - a$ is irreducible in $P(F)$ and K is generated over F by a root of $X^p - X - a$.*

Proof. Let τ be a generator of the cyclic Galois group G of K over F.

If e is the identity element of F, it is left fixed by all the elements of G; hence $S_{K/F}(e) = e+e+\ldots+e = pe = 0$. It follows from Theorem 23.1 that there exists an element α' of K such that $e = \alpha' - \tau(\alpha')$; if we set $\alpha = -\alpha'$ we have $\tau(\alpha) = \alpha + e$. Then, by induction, we deduce that for $k = 0, 1, \ldots, p-1$ we have $\tau^k(\alpha) = \alpha + ke$. Thus the subgroup of G which leaves α (and hence $F(\alpha)$) fixed consists of the identity automorphism alone. Hence, by Theorem 16.1, $(K: F(\alpha)) = 1$ and so $K = F(\alpha)$.

Now let us examine

$$\begin{aligned}\tau(\alpha^p - \alpha) = (\tau(\alpha))^p - \tau(\alpha) &= (\alpha+e)^p - (\alpha+e)\\ &= \alpha^p + e - \alpha - e = \alpha^p - \alpha;\end{aligned}$$

by induction, $\tau^k(\alpha^p - \alpha) = \alpha^p - \alpha$ for $k = 0, 1, \ldots, p-1$. Thus $\alpha^p - \alpha$ belongs to F (since F is the precise fixed field under G). Set $\alpha^p - \alpha = a$. This shows that α is a root of the polynomial $X^p - X - a$ in $P(F)$; hence the minimum polynomial m of α over F is a factor of $X^p - X - a$. But $\partial m = (F(\alpha): F) = p$ and m is a monic polynomial; so $m = X^p - X - a$ and hence $X^p - X - a$ is irreducible in $P(F)$.

It is clear from the form of the roots $\tau^k(\alpha) = \alpha + ke$ of $X^p - X - a$ that they all belong to K, which is therefore a splitting field over F for this polynomial.

THEOREM 23.5. *Let F be a field of non-zero characteristic p. Let a be a non-zero element of F and K a splitting field of $X^p - X - a$ over F including F. Then either $K = F$ or K is a cyclic extension of degree p over F generated over F by any one of the roots of $X^p - X - a$.*

Proof. Let α be any root of $X^p - X - a$ in K. Then $\alpha^p - \alpha = a$ and hence $(\alpha+e)^p - (\alpha+e) = \alpha^p + e - \alpha - e = a$

also. By induction it follows that $X^p - X - a$ has p distinct roots in K—namely α, $\alpha+e$, $\alpha+2e$, ..., $\alpha+(p-1)e$—all of which lie in $F(\alpha)$. Thus $K = F(\alpha)$.

Each monic irreducible factor of $X^p - X - a$ in $P(F)$ is the minimum polynomial over F of at least one of the roots of $X^p - X - a$. But since each root generates the same simple extension of F it follows that the minimum polynomials of all the roots have the same degree; hence all the irreducible factors of $X^p - X - a$ in $P(F)$ have the same degree, d say. Then p is a multiple of d and so, since p is a prime number, d is either 1 or p.

If $d = 1$ all the irreducible factors of $X^p - X - a$ in $P(F)$ are linear, and so all the roots of $X^p - X - a$ lie in F. In this case $K = F$.

On the other hand, if $d = p$, then $X^p - X - a$ is irreducible in $P(F)$. Since it has distinct roots, its splitting field K is a separable normal extension of degree p over F. Thus the Galois group G of K over F is of order p and hence is cyclic. Each automorphism in G is completely determined once its action on α is known; since there are p automorphisms and each must map α on another root of $X^p - X - a$, they are described by setting $\tau_k(\alpha) = \alpha + ke$ ($k = 0, 1, \ldots, p-1$), and clearly $\tau_k = \tau_1^k$, thus showing again that G is cyclic, generated by τ_1.

§ **24. Wedderburn's Theorem.** A **division ring** is a ring with identity which contains at least two elements and has the property that every non-zero element has an inverse relative to the multiplication. In the notation of § 1, a division ring is a ring satisfying the conditions **A1, A2, A3, A4, M1, M3, M4, AM**, that is to say, all the conditions for a field except the commutativity of multiplication. Division rings are also known as **skew fields.** A subring of any ring R which happens to be a division ring is called a **division subring** of R.

As in § 2 we may deduce that the product of any two

non-zero elements in a division ring D is non-zero and hence that the set of non-zero elements equipped with the multiplication operation forms a group; we denote this group by D^* and call it the **multiplicative group** of the division ring.

The **centre** of a division ring D is the subset Z of D consisting of all the elements of D which commute with every element of D; that is to say, an element z of D belongs to Z if and only if $zx = xz$ for every element x of D. Clearly the zero element of D belongs to Z; if we denote by Z^* the set of non-zero elements of Z it follows that Z^* is the centre of the multiplicative group D^*.

Let a be any element of a division ring D. Then the **normaliser** of a in D is the set $N(a)$ consisting of elements of D which commute with a : so n belongs to $N(a)$ if and only if $an = na$.

THEOREM 24.1. *Let D be a division ring. Then the centre Z of D is a subfield of D and the normaliser of each element of D is a division subring of D including Z.*

Proof. Let z_1 and z_2 be elements of Z; let x be any element of D. Then $xz_1 = z_1x$ and $xz_2 = z_2x$, whence

$$x(z_1+z_2) = xz_1+xz_2 = z_1x+z_2x = (z_1+z_2)x$$

and

$$x(z_1z_2) = (xz_1)z_2 = (z_1x)z_2 = z_1(xz_2) = z_1(z_2x) = (z_1z_2)x.$$

Further, if z is an element of Z and x is any element of D, we have

$$x(-z) = -(xz) = -(zx) = (-z)x,$$

and if z is non-zero we have

$$xz^{-1} = z^{-1}(zx)z^{-1} = z^{-1}(xz)z^{-1} = z^{-1}x.$$

Hence Z is a division subring of D. But since the elements of Z commute with all the elements of D, and hence in particular with one another, Z is in fact a field.

The proof of the second part of the theorem is similar to that of the first.

Now let D be a division ring with a finite number of elements. Then the centre Z of D is a finite field, with q elements, say. If D_1 is any division subring of D containing Z, D_1 may be regarded as a vector space over Z—the operations of the vector space being the ordinary addition in D_1 and the multiplication (in D_1) of elements of D_1 by elements of Z. If the dimension of this vector space over Z is d_1 it follows that D_1 has q^{d_1} elements. In particular, there is an integer n such that D itself contains q^n elements.

Now let C be any class of conjugate elements in D^*, the multiplicative group of D. If x is any element of C, the number of elements in the class C is equal to the index in D^* of the normaliser of x in D^*. The normaliser $N(x)$ of x in D is a division subring of D including Z (Theorem 24.1) and so there is an integer r_C such that $N(x)$ contains q^{r_C} elements. It is easy to see that the normaliser of x in D^* is just the set of non-zero elements of $N(x)$ and so consists of $q^{r_C}-1$ elements. It follows that the class C consists of $(q^n-1)/(q^{r_C}-1)$ elements; and since this number is of course an integer we deduce that r_C is a factor of n.

We are now in a position to establish Wedderburn's important theorem on finite division rings.

THEOREM 24.2. *Every finite division ring is a field.*

Proof. Let D be a finite division ring, with centre Z. Suppose Z has q elements and D has q^n elements. We want to prove that every element of D commutes with every other element, i.e., that $D = Z$ and $n = 1$.

We split up the multiplicative group D^* into its conjugate classes. Each element of Z forms a conjugate class by itself; suppose there are, in addition, classes $C_1, \ldots, C_k$ each containing more than one element. Then, as we saw above, corresponding to each class C_i there is a factor r_i of n such that C_i consists of $(q^n-1)/(q^{r_i}-1)$ elements, and of course $r_i<n$. Hence, counting up the elements in the various classes, we obtain

$$q^n-1 = q-1+\sum_{i=1}^{k}\frac{q^n-1}{q^{r_i}-1}. \tag{24.1}$$

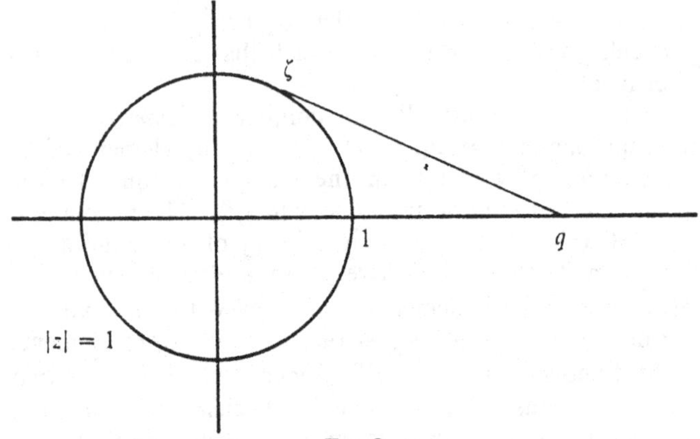

FIG. 8

Now the nth cyclotomic polynomial Φ_n in $P(\mathbf{Q})$ is a factor of X^n-1 and of each polynomial $(X^n-1)/(X^{r_i}-1)$ for which r_i is a factor of n less than n. Let a be the integer we obtain on substituting the integer q in the polynomial Φ_n (which has integer coefficients by Theorem 21.2). Then a is a factor of q^n-1 and of each term $(q^n-1)/(q^{r_i}-1)$ in the summation on the right of (24.1). Hence a is a factor of $q-1$.

If $n>1$, then for every primitive nth root of unity ζ in the field of complex numbers $\mathbf{C}$ we have $|q-\zeta|>q-1$ (see fig. 8). Hence $|a| = \prod |q-\zeta| > q-1$ (where the product is extended over all the primitive roots of unity), and hence a cannot be a factor of $q-1$.

It follows that there are no classes C_i containing more than one element. Hence $n = 1$ and $D = Z$, as required.

§ **25. Ruler-and-compasses constructions.** Among the problems which were studied by the geometers of antiquity were three which achieved great notoriety and which even to this day attract attention from enthusiastic, but misguided, amateur mathematicians. These problems are: first, that of trisecting an arbitrary given angle; second, that of constructing a cube whose volume is double that of a given cube (" duplication of the cube "); third, that of constructing a square whose area is equal to that of a given circle (" squaring the circle "). In each case the only tools allowed are the traditional geometrical instruments—ruler and compasses. We shall show in this section that these three problems are insolvable.

First we must make quite precise what we mean by a " ruler-and-compasses construction ". So we consider the Euclidean plane and recall from elementary coordinate geometry that if we specify in this plane two straight lines at right angles meeting at a point O and also a point I on one of those lines then we can set up a Cartesian coordinate system in the plane by taking O as origin and I to be the point (1, 0). Let $\mathscr{B}$ be a collection of points in this plane, including O and I; intuitively we think of the points in $\mathscr{B}$ as " given " or " known " points from which to start our construction. The points in this collection $\mathscr{B}$ will be called **basic points.**

Then a **ruler-and-compasses construction based on** $\mathscr{B}$ is a finite sequence of operations of the following types:

(1) drawing a straight line through two points which are either basic points or points previously constructed in the sequence of operations;

(2) drawing a circle with centre at a basic point or a point previously constructed, and with radius equal to the distance between two points, each of which is either a basic point or a point previously constructed;

(3) marking the points of intersection of (*a*) pairs of straight lines, (*b*) pairs of circles, (*c*) straight lines and circles constructed in (1) and (2).

Any point P which is obtained by an operation of type (3) in a ruler-and-compasses construction based on $\mathscr{B}$ is said to be **constructible from $\mathscr{B}$.** If $\mathscr{B}$ consists of the points O and I and no others, we simply say that P is **constructible.**

The coordinates of the points of $\mathscr{B}$ in the Cartesian coordinate system introduced into the plane by means of O and I are real numbers; let $\boldsymbol{B}$ be the subfield of the real number field $\mathbf{R}$ obtained by adjoining all these coordinates to the rational number field $\mathbf{Q}$. (Of course, if O and I are the only basic points, we have $\boldsymbol{B} = \mathbf{Q}$.)

Let now P be any point of the plane with coordinates (α, β) in the coordinate system determined by O and I. The subfield of $\mathbf{R}$ obtained by adjoining α and β to $\boldsymbol{B}$ will be denoted by $\boldsymbol{B}(P)$. We now proceed to give, in terms of this subfield of $\mathbf{R}$, a necessary condition that the point P be constructible from $\mathscr{B}$.

THEOREM 25.1. *If the point P is constructible from $\mathscr{B}$, then the degree of $\boldsymbol{B}(P)$ over $\boldsymbol{B}$ is a power of 2.*

Proof. Let $P_1, P_2, \ldots, P_n = P$ be the sequence of points obtained by operations of type (3) in a ruler-and-compasses construction of P from $\mathscr{B}$; we may suppose that P_1 is one of the basic points. Let (α_i, β_i) be the coordinates of P_i $(i = 1, \ldots, n)$.

Let K be the subfield of $\mathbf{R}$ obtained by adjoining to $\boldsymbol{B}$ the coordinates of $P_1, \ldots, P_n$; we shall show that the degree of K over $\boldsymbol{B}$ is a power of 2. The desired result concerning $B(P)$ will then follow at once, since $B(P)$ is a subfield of K and hence $(B(P): B)$ is a factor of $(K: B)$. We proceed by induction on n.

If $n = 1$ we have $K = B(P_1)$, and since P_1 is a basic point we have $B(P_1) = B$; thus $(K: B) = 1 = 2^0$.

Now suppose we have established that if K_{k-1} is the subfield of $\mathbf{R}$ obtained by adjoining to B the coordinates of $P_1, \ldots, P_{k-1}$ then $(K_{k-1}: B)$ is a power of 2.

It follows from elementary coordinate geometry that if P_i and P_j are distinct points $(1 \leqq i, j \leqq k-1)$ then the straight line λ_{ij} joining them is represented by a linear equation with coefficients in K_{k-1}, namely

$$(\alpha_j - \alpha_i)(y - \beta_i) = (\beta_j - \beta_i)(x - \alpha_i).$$

Similarly, if P_r and P_s are distinct points and P_t is any point $(1 \leqq r, s, t \leqq k-1)$, the circle $\sum_{rs}^{t}$, with centre P_t and radius equal to the distance between P_r and P_s is represented by a quadratic equation with coefficients in K_{k-1}, namely

$$(x - \alpha_t)^2 + (y - \beta_t)^2 = (\alpha_r - \alpha_s)^2 + (\beta_r - \beta_s)^2. \qquad (25.1)$$

Let K_k be the subfield of $\mathbf{R}$ obtained by adjoining to B the coordinates of $P_1, \ldots, P_k$; then of course K_k may also be obtained by adjoining to K_{k-1} the coordinates of P_k.

If P_k is obtained from $P_1, \ldots, P_{k-1}$ by an operation of type 3(a), i.e., if it is the point of intersection of two lines like λ_{ij}, then its coordinates are obtained by solving simultaneously two linear equations with coefficients in P_{k-1}. It follows that the coordinates of P_k lie in K_{k-1}; thus $K_k = K_{k-1}$ and so $(K_k: B) = (K_{k-1}: B)$ is a power of 2.

Next suppose P_k is obtained from $P_1, \ldots, P_{k-1}$ by an operation of type 3(c), i.e., that P_k is one of the points of intersection of a line like λ_{ij} and a circle like $\sum_{rs}^{t}$. In this case the coordinates of P_k are obtained by solving simultaneously a linear and a quadratic equation with coefficients in K_{k-1}. We recall from elementary algebra the procedure adopted for solving such a pair of equations: we use the linear equation to express one of the unknowns —say y—in terms of the other, and substitute this expression in the quadratic equation; we thus obtain a quadratic equation for x with coefficients in K_{k-1}. The roots of this equation, one of which is α_k, lie either in K_{k-1} itself or in an extension L of degree 2 over K_{k-1}; since β_k is expressed linearly in terms of α_k with coefficients in K_{k-1}, it follows that β_k also lies either in K_{k-1} or in L. Thus $(K_k: K_{k-1})$ is either 1 or 2, and hence $(K_k: \boldsymbol{B})$ is a power of 2.

Finally, suppose P_k is obtained by an operation of type 3(b), i.e., that P_k is one of the points of intersection of two circles like $\sum_{rs}^{t}$. Then the coordinates of P_k are obtained by solving simultaneously two quadratic equations of type (25.1) with coefficients in K_{k-1}. But the simultaneous solutions of two such quadratic equations are precisely the same as the simultaneous solutions of one of them and the linear equation obtained by subtracting the quadratics. The discussion of the previous paragraph shows that in this case also $(K_k: \boldsymbol{B})$ is a power of 2.

This completes the induction and the theorem is established.

We can now dispose of two of the problems of antiquity mentioned at the beginning of the section. First of all it is easy to see that the " duplication of the cube " is equivalent to the construction from the basic points O and I of the point $(\alpha, 0)$, where α is the real number such that $\alpha^3 = 2$. Since the polynomial X^3-2 is irreducible in $P(\mathbf{Q})$ (§ 10, Example 1), the field $\mathbf{Q}(\alpha)$ has degree 3 over $\mathbf{Q}$ and hence, since 3 is not a power of 2, the point $(\alpha, 0)$ is not constructible from O and I. Next, " squaring the circle " is equivalent to the construction (again from the two basic points O and I) of the point $(\sqrt{\pi}, 0)$, where π (for the first time in this book) denotes the ratio of the circumference of a circle to its diameter. But in 1882

Lindemann proved that π is not algebraic over the field of rational numbers. Hence $(\mathbf{Q}(\pi): \mathbf{Q})$ is not even finite, far less a power of 2. It follows that $(\mathbf{Q}(\sqrt{\pi}): \mathbf{Q})$ is not a power of 2 and hence $(\sqrt{\pi}, 0)$ is not constructible.

In order to give a sufficient condition for the constructibility of a point from $\mathscr{B}$ we must recall informally some of the ruler-and-compasses constructions of elementary Euclidean geometry. We shall not go into great detail over these, leaving it to the reader, if he feels inclined, to express the constructions formally as sequences of operations of our three types.

First of all, it is easy to see how one may construct from O and I for every integer m and every non-zero integer n the point with coordinates $(m/n, 0)$. Namely, we mark $n+1$ points $P_0 = O, P_1, \ldots, P_n$ on the Y-axis such that $P_iP_{i+1} = OI\,(i = 0, 1, \ldots, n-1)$; we join P_n to I and draw through P_1 a line parallel to P_nI meeting the X-axis in Q; then Q is the point $(1/n, 0)$ and it is then a simple matter to obtain the points $(m/n, 0)$ for all integers m (see fig. 9).

Next, if P is any point in $\mathscr{B}$ with coordinates (ξ, η), it is an easy matter to construct from $\mathscr{B}$ the points with coordinates $(\xi, 0)$ and $(\eta, 0)$.

Now suppose we have constructed from $\mathscr{B}$ the points $A = (a, 0)$ and $B = (b, 0)$. Simple constructions allow us to obtain the points (a, b), $(a+b, 0)$, $(a-b, 0)$; we shall not take space to describe them. The construction of $(ab, 0)$ is perhaps a little less familiar: namely, mark B' on the Y-axis such that $OB' = OB$; join IB' and draw a line through A parallel to IB' meeting the Y-axis in C'; mark C on the X-axis such that $OC = OC'$, taking C on the positive axis if a and b have like sign and on the negative axis if a and b have unlike sign; then a simple argument using similar triangles shows that $C = (ab, 0)$ (see fig. 10). Similarly, if b is non-zero, we can construct $(a/b, 0)$.

This discussion shows first that all the points with coordinates in the basic field $\boldsymbol{B}$ can be constructed from $\mathscr{B}$ and then that if the point with coordinates $(\alpha, 0)$ is constructible from $\mathscr{B}$ so also are all the points with coordinates in the field $\boldsymbol{B}(\alpha)$.

But if α is positive and the point D with coordinates $(\alpha, 0)$ is constructible from $\mathscr{B}$ there is also a ruler-and-compasses construction which yields the point $(\sqrt{\alpha}, 0)$. We mark D_1 to the right of I such that $ID_1 = OD$; draw a circle on OD_1 as diameter, cutting the ordinate through I at B; mark C on the X-axis such that $OC = IB$. Then C is the point $(\sqrt{\alpha}, 0)$ (see fig. 11).

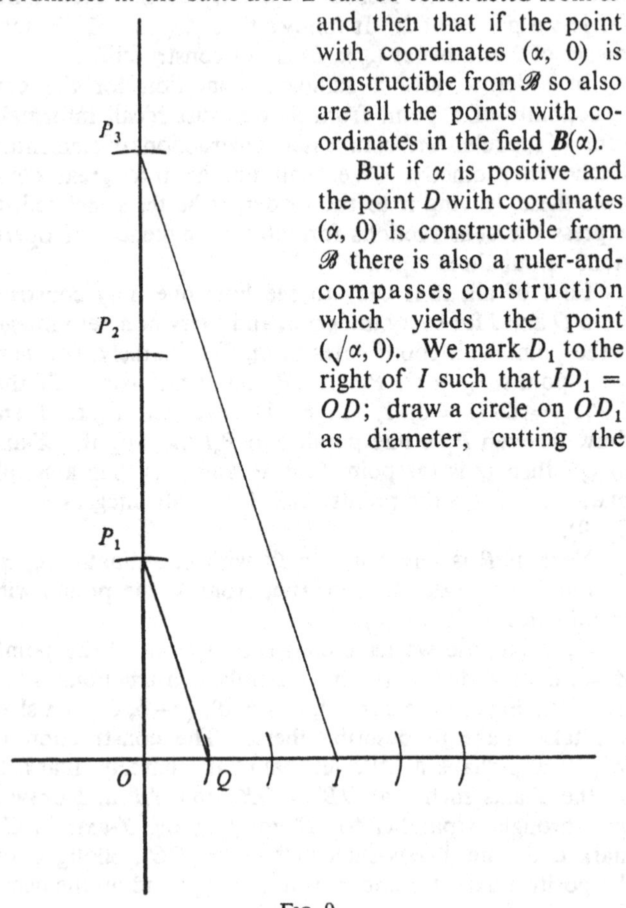

FIG. 9

We are now ready to give a sufficient condition for constructibility.

THEOREM 25.2. *Let P be a point in the plane. If the field* $\boldsymbol{B}(P)$ *has a sequence of subfields* $\boldsymbol{B}(P) = K_n, K_{n-1}, \ldots, K_1, K_0 = \boldsymbol{B}$ *such that* K_i *includes* K_{i-1} *and* $(K_i : K_{i-1}) = 2$ $(i = 1, \ldots, n)$, *then the point P is constructible from* $\mathscr{B}$.

Proof. We proceed by induction on n.

If $n = 0$ then $\boldsymbol{B}(P) = \boldsymbol{B}$. Hence P is constructible from $\mathscr{B}$ according to the preceding discussion.

Now let K be a field which has a sequence of subfields $K = K_k$, $K_{k-1}, \ldots, K_1, K_0 = \boldsymbol{B}$ such that K_i includes K_{i-1} and $(K_i : K_{i-1}) = 2$ $(i = 1, \ldots, k)$ and suppose we have established that every point with coordinates in K_{k-1} is constructible from $\mathscr{B}$.

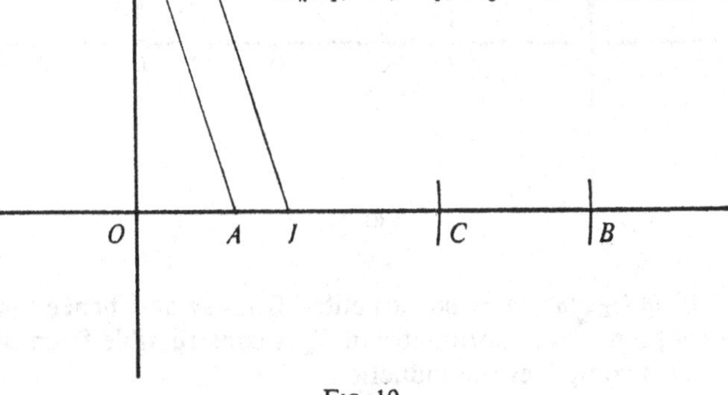

FIG. 10

Since $(K_k : K_{k-1}) = 2$, it follows that if β is any element of K_k which does not lie in K_{k-1} then $K_k = K_{k-1}(\beta)$. Suppose the minimum polynomial of β over K_{k-1} is

$X^2+aX+b = (X+\frac{1}{2}a)^2+(b-\frac{1}{4}a^2)$. If we set $\alpha = \beta+\frac{1}{2}a$ we have $\alpha^2 = \frac{1}{4}a^2-b\geqq 0$; thus α^2 is a positive element of K_{k-1} and clearly $K_k = K_{k-1}(\beta) = K_{k-1}(\alpha)$.

Now, since the point $(\alpha^2, 0)$ has coordinates in K_{k-1}, it is constructible from $\mathscr{B}$ by the inductive hypothesis. The discussion preceding this theorem then shows that

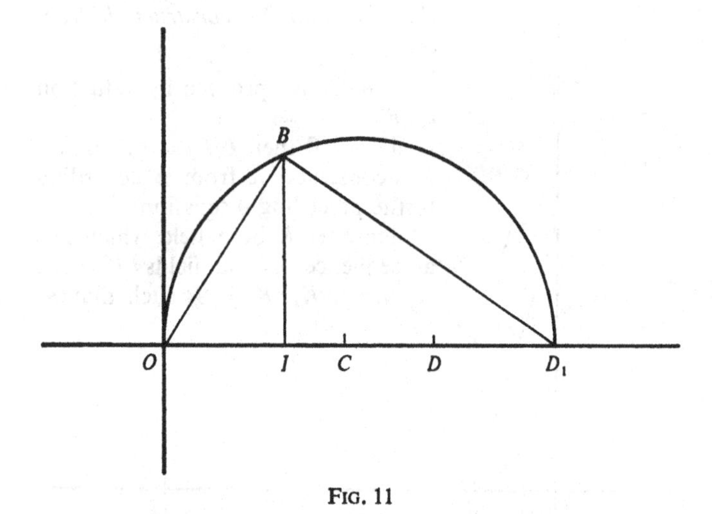

FIG. 11

$(\alpha, 0) = (\pm\sqrt{\alpha^2}, 0)$ is constructible from $\mathscr{B}$ and hence that every point with coordinates in K_k is constructible from $\mathscr{B}$.
This completes the induction.

COROLLARY. *Let P be a point in the plane. If the field* $\boldsymbol{B}(P)$ *is a normal extension of* $\boldsymbol{B}$ *such that* $(\boldsymbol{B}(P): \boldsymbol{B})$ *is a power of* 2, *then the point P is constructible from* $\mathscr{B}$.

Proof. Let G be the Galois group of $\boldsymbol{B}(P)$ over $\boldsymbol{B}$; the order of G is a power of 2, say 2^s. Then, according to the theory of groups of prime power order, G has a sequence of subgroups, $G = A_0, A_1, \ldots, A_s = \{\varepsilon\}$, each normal and of index 2 in the preceding. It follows from Theorem 16.3 that $\boldsymbol{B}(P)$ has a sequence of subfields $\boldsymbol{B}(P) = K_s, K_{s-1}, \ldots, K_0 = \boldsymbol{B}$ each of degree 2 over the next. Hence P is constructible from $\mathscr{B}$ by Theorem 25.2.

Let now AOI be a given angle; set $\cos AOI = a$, $\cos \frac{1}{3}AOI = \alpha$. Once AOI is given, the point J with coordinates $(a, 0)$ is easily constructed. The problem of trisecting the angle AOI is now clearly equivalent to that of constructing the point $(\alpha, 0)$ from the basic set $\mathscr{B} = \{O, I, J\}$. The basic field $\boldsymbol{B}$ in this case is the simple extension $\mathbf{Q}(a)$ of $\mathbf{Q}$.

Since for every angle θ we have

$$\cos 3\theta = 4\cos^3\theta - 3\cos\theta,$$

it follows that α is a root of the polynomial $f = 4X^3 - 3X - a$ with coefficients in $\boldsymbol{B}$. If this polynomial f is reducible in $P(\boldsymbol{B})$ then the degree $(\boldsymbol{B}(\alpha): \boldsymbol{B})$ is either 1 or 2 and hence $(\alpha, 0)$ is constructible from $\mathscr{B}$ by Theorem 25.2. But if f is irreducible in $P(\boldsymbol{B})$, then $(\boldsymbol{B}(\alpha): \boldsymbol{B}) = 3$ and so Theorem 25.1 shows that $(\alpha, 0)$ is not constructible from $\mathscr{B}$.

These remarks show that the angle AOI can be trisected by a ruler-and-compasses construction if and only if the polynomial $4X^3 - 3X - a$ is reducible in $P(\mathbf{Q}(a))$. For example, a right angle can be trisected by ruler and compasses, but an angle of 60° cannot.

Another type of construction problem which interested the Greeks was that of constructing regular polygons using ruler-and-compasses constructions. Euclid gave such constructions for an equilateral triangle, a square, a regular pentagon, a regular hexagon and a regular polygon of fifteen sides. We shall now determine for which positive integers n regular n-sided figures can be constructed by ruler-and-compasses constructions. The basic points for this investigation will be simply O and I, and the basic field $\boldsymbol{B}$ is thus the field of rational numbers $\mathbf{Q}$.

Let θ_n be the angle subtended at the centre of a circle by one side of a regular n-sided figure inscribed in it. Then it is clear that the problem of constructing a regular polygon of n sides, $IA_1A_2...A_{n-1}$ is equivalent to that of constructing the point C_n with coordinates $(\cos \theta_n, 0)$ (see fig. 12).

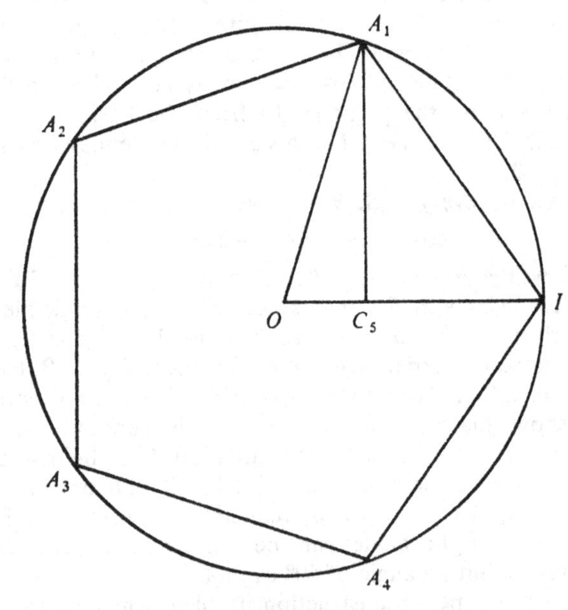

FIG. 12

We remark that if θ is any angle such that $(\cos \theta, 0)$ is constructible, then $(\sin \theta, 0)$ is also constructible, and hence so are $(\cos k\theta, 0)$ and $(\sin k\theta, 0)$ for all integers k. Hence, if $(\cos \theta, 0)$ and $(\cos \phi, 0)$ are constructible, then so are $(\cos (r\theta + s\phi), 0)$ for all integers r and s.

Now let m and n be integers such that a regular polygon of mn sides is constructible. Then $(\cos \theta_{mn}, 0)$ is constructible, and hence, since $\theta_n = m\theta_{mn}$ and $\theta_m = n\theta_{mn}$, the points $(\cos\theta_n, 0)$ and $(\cos\theta_m, 0)$ are constructible; thus regular polygons of n and m sides are constructible.

Conversely, suppose that regular polygons of m and n sides are constructible, where m and n are relatively prime. Then there are integers r and s such that $rm+sn = 1$, whence $r\theta_n+s\theta_m = \theta_{mn}$. Since $(\cos\theta_n, 0)$ and $(\cos\theta_m, 0)$ are constructible, so is $(\cos\theta_{mn}, 0)$ and hence a regular polygon of mn sides is constructible.

THEOREM 25.3. *A regular polygon of n sides can be constructed by a ruler-and-compasses construction if and only if* $n = 2^k p_1 \ldots p_r$ *where k is any non-negative integer and* $p_1, \ldots, p_r$ *are distinct prime numbers of the form* $2^{2^m}+1$.

Proof. (1) Suppose that a regular polygon of n sides is constructible.

Let $n = p_1^{k_1} \ldots p_s^{k_s}$ be the expression of n as a product of powers of distinct primes. Then, according to the remarks preceding the theorem, regular polygons of $p_j^{k_j}$ sides are constructible ($j = 1, \ldots, s$). If q is any one of these prime powers $p_j^{k_j}$, it follows that $(\cos\theta_q, 0)$ is constructible and hence, according to Theorem 25.1, the degree $(\mathbf{Q}(\cos\theta_q): \mathbf{Q})$ is a power of 2.

Now let $\zeta_q = \exp(i\theta_q)$; ζ_q is a primitive qth root of unity in the complex number field. Further,

$$\zeta_q+\zeta_q^{-1} = 2\cos\theta_q,$$

so $\zeta_q^2-(2\cos\theta_q)\zeta_q+1 = 0$; hence $(\mathbf{Q}(\zeta_q): \mathbf{Q}(\cos\theta_q))$ is either 1 or 2. It follows that $(\mathbf{Q}(\zeta_q): \mathbf{Q})$ is a power of 2. According to Theorem 22.3,

$$(\mathbf{Q}(\zeta_q): \mathbf{Q}) = \phi(q) = \phi(p_j^{k_j}) = p_j^{k_j-1}(p_j-1).$$

Since $p_j^{k_j-1}(p_j-1)$ is a power of 2, there are only two possibilities: either $p_j = 2$ and k_j is unrestricted, or $k_j = 1$ and p_j-1 is a power of 2. Now we consider the integers of the form $2^\alpha+1$, and show that if α is not itself a power of 2 then $2^\alpha+1$ is not a prime number. For suppose $\alpha = \lambda\mu$, where λ is an odd number; then $X^\lambda+1$ is divisible by $X+1$ and hence $(2^\mu)^\lambda+1 = 2^\alpha+1$ is divisible by $2^\mu+1$.

Thus, if $n = p_1^{k_1}...p_s^{k_s}$ and a regular polygon of n sides is constructible, then the prime power factors of n are either 2^k, with k unrestricted, or else single prime numbers of the form $2^{2^m}+1$.

(2) Conversely, suppose n is a product of such prime power factors.

For each factor $q = p_j^{k_j}$ we see that $\phi(q)$ is a power of 2 and hence $(\mathbf{Q}(\zeta_q): \mathbf{Q})$ is a power of 2. Now $\mathbf{Q}(\zeta_q)$ is a separable normal extension of $\mathbf{Q}$, and its Galois group, which is isomorphic to the group of relatively prime residue classes modulo q (Theorem 22.3), is abelian; thus all the subgroups of the Galois group are normal and hence all the subfields of $\mathbf{Q}(\zeta_q)$ are normal extensions of $\mathbf{Q}$, by Theorem 16.3.

In particular, $\mathbf{Q}(\cos\theta_q)$ is a normal extension of $\mathbf{Q}$ whose degree over $\mathbf{Q}$ is a power of 2. It follows from the Corollary to Theorem 25.2 that $(\cos\theta_q, 0)$ is constructible. Hence each regular polygon of $p_j^{k_j}$ sides is constructible and so a regular polygon of n sides is constructible.

The prime numbers of the form $2^{2^m}+1$ are called **Fermat primes**; the only known Fermat primes are 3, 5, 17, 257 and 65537.

§ **26. Solution by radicals.** Throughout this section we consider only fields of characteristic zero.

Let F be a field of characteristic zero; a field E containing F is said to be an **extension** of F **by radicals** if there exists a sequence of subfields $F = E_0, E_1, ..., E_{r-1}, E_r = E$ such that for $i = 0, ..., r-1$, $E_{i+1} = E_i(\alpha_i)$, where α_i is a root of an irreducible polynomial in $P(E_i)$ of the form $X^{n_i}-a_i$. If E is an extension of F by radicals, the elements of E can thus be obtained from those of F by means of the field operations together with the extraction of n_ith roots ($i = 1, \ldots, r$). A polynomial f in $P(F)$ is said to be **solvable by radicals** if there exists a splitting field of f over F which is included in an extension of F by radicals. We shall establish in Theorems 26.4 and 26.5 a necessary and sufficient condition that a polynomial be solvable by radicals.

We begin with a number of preliminary results.

THEOREM 26.1. *Let F be a field of characteristic zero, K a normal extension of F including F with an abelian Galois group. If $(K: F) = n$ and the polynomial $k_n = X^n - e$ splits completely in $P(F)$, then K is an extension of F by radicals.*

Proof. Let G be the abelian Galois group of K over F. According to the theory of abelian groups, G may be expressed as a direct product of cyclic groups, say

$$G = Z_1 \times \ldots \times Z_r.$$

For $i = 0, \ldots, r-1$ set $G_i = Z_1 \times \ldots \times Z_{r-i}$, and let G_r be the subgroup of G consisting of the identity element alone; then G_{i+1} is a normal subgroup of G_i with factor group isomorphic to the cyclic group Z_{r-i} ($i = 0, \ldots, r-1$).

Let E_i be the subfield of K left fixed by G_i ($i = 0, \ldots, r$). According to the Corollary to Theorem 16.3, E_{i+1} is a normal extension of E_i with cyclic Galois group, isomorphic to Z_{r-i} ($i = 0, \ldots, r-1$). Since the degree n_i of E_{i+1} over E_i is a factor of n, and k_n splits completely in $P(F)$—and hence in $P(E_i)$—it follows that k_{n_i} splits completely in $P(E_i)$. So we may apply Theorem 23.2 and deduce that $E_{i+1} = E_i(\alpha_i)$ where α_i is a root of an irreducible polynomial in $P(E_i)$ of the form $X^{n_i} - a_i$ ($i = 0, \ldots, r-1$).

Thus K is an extension of F by radicals, as asserted.

THEOREM 26.2. *Let F be a field of characteristic zero. For every positive integer n, the polynomial $k_n = X^n - e$ in $P(F)$ is solvable by radicals.*

Proof. We proceed by induction on n.

If $n = 1$, F is itself a splitting field for k_n over F; since F is also an extension of F by radicals, the required result follows at once.

Now suppose we have established for all integers l less than m that the polynomial k_l is solvable by radicals.

Let K_m be a splitting field of k_m over F including F; if $(K_m: F) = r$ we may deduce from Theorem 21.1 that $r \leqq \phi(m) < m$. According to the inductive hypothesis, there is a splitting field K_r of k_r over F which is included in an extension E of F by radicals. According to Theorem 12.3 we may assume without loss of generality that E is included in the same algebraic closure C of F as K_m; then we may form the compositum $L = E(K_m)$ of E and K_m in C. The relations between the fields involved in this argument are illustrated in fig. 13.

Theorem 18.3 shows that L is a separable normal extension of E and that the Galois group H of L over E is isomorphic to a subgroup of the Galois group of K_m over F. Hence H is abelian (by Theorem 21.1). It follows also that the degree s of L over E is a factor of $r = (K_m: F)$. Since k_r splits completely in $P(E)$, so too does k_s. Thus we may apply Theorem 26.1 and deduce that L is an extension of E by radicals. Since E is already an extension of F by radicals it follows that L is also an extension of F by radicals, and hence k_m is solvable by radicals.

This completes the induction.

Let G be a finite group, with identity element ε; G is said to be a **solvable** group if there exists a sequence of subgroups

$$G = G_0, G_1, \ldots, G_{s-1}, G_s = \{\varepsilon\}, \qquad (26.1)$$

such that for $i = 1, \ldots, s$, G_i is a normal subgroup of G_{i-1} with an abelian factor group. It is clear at once that all abelian groups are solvable. Since each of the abelian factor groups in (26.1) is isomorphic to a direct product of cyclic groups of prime power order and each of these cyclic groups has a sequence of subgroups each normal and of prime index in the preceding, we may insert additional subgroups in the sequence (26.1) and obtain a sequence of

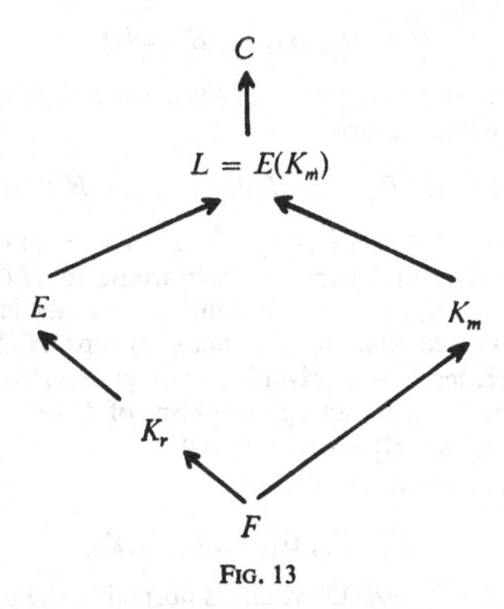

FIG. 13

subgroups, each normal in the preceding and such that the factor groups are cyclic and of prime order. We note for future reference the classical result that for $n > 4$ the symmetric group S_n on n digits is not solvable.

We now prove two elementary properties of solvable groups.

THEOREM 26.3. *All the subgroups and all the epimorphic images of a solvable group are solvable.*

Proof. Let G be a solvable group and let

$$G = G_0, G_1, \ldots, G_s = \{\varepsilon\}$$

be a sequence of subgroups of G, each normal in the preceding and such that all the factor groups G_i/G_{i+1} are abelian ($i = 0, \ldots, s-1$).

(1) First let H be a subgroup of G. Set $H_i = H \cap G_i$ ($i = 0, \ldots, s$) and consider the sequence of subgroups of H:

$$H = H_0, H_1, \ldots, H_s = \{\varepsilon\}.$$

It is easy to verify that each of these subgroups is normal in the preceding. Since

$$H_{i+1} = H \cap G_{i+1} = H \cap G_i \cap G_{i+1} = H_i \cap G_{i+1}$$

it follows that the factor group H_i/H_{i+1} may be expressed as $H_i/H_i \cap G_{i+1}$ and hence is isomorphic to H_iG_{i+1}/G_{i+1}. Since H_iG_{i+1}/G_{i+1} is a subgroup of the abelian group G_i/G_{i+1}, we see that all the factor groups H_i/H_{i+1} are abelian. Hence H is a solvable group, as asserted.

(2) Now let ϕ be an epimorphism of G onto a group G'. If we set $G'_i = \phi(G_i)$ ($i = 0, \ldots, s$) we obtain a sequence of subgroups of G':

$$G' = G'_0, G'_1, \ldots, G'_s = \{\varepsilon'\}.$$

Clearly each of these subgroups is normal in the preceding. In order to show that the factor groups G'_i/G'_{i+1} are all abelian, we define mappings ϕ_i^* of G_i/G_{i+1} into G'_i/G'_{i+1} as follows. Let C be a coset of G_i relative to G_{i+1}; if τ is any element of C we define $\phi_i^*(C)$ to be the coset of $\phi(\tau)$ relative to G'_{i+1}. It is easily verified that $\phi_i^*(C)$ does not depend upon the choice of the element τ from C, and that ϕ_i^* is an epimorphism of G_i/G_{i+1} onto G'_i/G'_{i+1}. It follows that the factor groups G'_i/G'_{i+1} are all abelian; so G' is solvable, as we claimed.

We are now ready to establish a criterion for a polynomial to be solvable by radicals.

THEOREM 26.4. *Let F be a field of characteristic zero. If a polynomial f in $P(F)$ has a splitting field over F with a solvable Galois group, then f is solvable by radicals.*

Proof. Let K be a splitting field of f over F including F, with solvable Galois group G.

If G has order n, let K_n be a splitting field of $k_n = X^n - e$ over F; according to Theorem 26.2, K_n is included in an extension of F by radicals, E say. We may assume without loss of generality that E is included in the same algebraic closure C of F as K (see Theorem 12.3). Thus we may form the compositum $L = E(K)$ of E and K in C. According to Theorem 18.3, L is a separable normal extension of E whose Galois group H is isomorphic to a certain subgroup of G; thus H is solvable, by Theorem 26.3. Let

$$H = H_0, H_1, \ldots, H_s = \{\varepsilon\}$$

be a sequence of subgroups of H, each normal in the preceding, such that the factor groups H_i/H_{i+1} are all abelian ($i = 0, \ldots, s-1$).

Let L_i be the subfield of L left fixed by H_i ($i = 0, \ldots, s$); then, according to the Corollary of Theorem 16.3, L_{i+1} is a normal extension of L_i whose Galois group is isomorphic to the abelian group H_i/H_{i+1} ($i = 0, \ldots, s-1$). Since the degree $n_i = (L_{i+1}: L_i)$ is a factor of $(L: E)$ and hence of $(K: F) = n$, and since the polynomial k_n splits completely in $P(E)$, it follows that k_{n_i} also splits completely in $P(E)$ and *a fortiori* in $P(L_i)$. Thus we may apply Theorem 26.1 and deduce that L_{i+1} is an extension of L_i by radicals ($i = 0, \ldots, s-1$). Hence, since E is an extension of F by radicals, it follows that L is an extension of F by radicals.

Thus f is solvable by radicals.

We now show that the sufficient condition which we

have just established for a polynomial to be solvable by radicals is also a necessary condition.

THEOREM 26.5. *Let F be a field of characteristic zero. If the polynomial f in $P(F)$ is solvable by radicals, then the Galois group of any splitting field of f over F is solvable.*

Proof. Let K be a splitting field of f over F, included in an extension of F by radicals, E say. Then there is a sequence of subfields of E,

$$F = E_0, E_1, \ldots, E_r = E,$$

such that $E_{i+1} = E_i(\alpha_i)$, where α_i is a root of a polynomial $X^{n_i} - a_i$, irreducible in $P(E_i)$ $(i = 0, \ldots, r-1)$.

Set $n = n_0 n_1 \ldots n_{r-1}$ and let K_n be a splitting field of $k_n = X^n - e$ over F, which we may assume as usual to lie in the same algebraic closure A of F as E. We form the compositum $K_n(E)$ and the normal closure N of this compositum over F in A; let $\bar{G}$ be the Galois group of N over F.

The field N has a sequence of subfields

$$N, K_n(E_r), K_n(E_{r-1}), \ldots, K_n(E_1), K_n(E_0) = K_n, F \quad (26.2)$$

each included in the preceding (see fig. 14). Let G_i be the subgroup of $\bar{G}$ which leaves $K_n(E_i)$ fixed $(i = 0, \ldots, r)$. Since n_i is a factor of n and k_n splits completely in $P(K_n)$ it follows that k_{n_i} splits completely in $P(K_n)$ and hence in $P(K_n(E_i))$. Thus $K_n(E_{i+1}) = (K_n(E_i))(\alpha_i)$ is a splitting field of $X^{n_i} - a_i$ over $K_n(E_i)$. It follows from Theorem 23.3 that $K_n(E_{i+1})$ is a normal extension of $K_n(E_i)$ with a cyclic Galois group $(i = 0, \ldots, r-1)$. We deduce from the Corollary to Theorem 16.3 that G_{i+1} is a normal subgroup of G_i and that the factor group G_i/G_{i+1} is cyclic $(i = 0, \ldots, r-1)$.

We now examine the sequence of subfields of K which we obtain by taking the intersections of K with the fields in the sequence (26.2):

$$K = K \cap K_n(E_r),\ K \cap K_n(E_{r-1}),\ \ldots,\ K \cap K_n(E_0) = K \cap K_n,\ F. \tag{26.3}$$

Since K is a normal extension of F, the subgroup G of $\bar{G}$

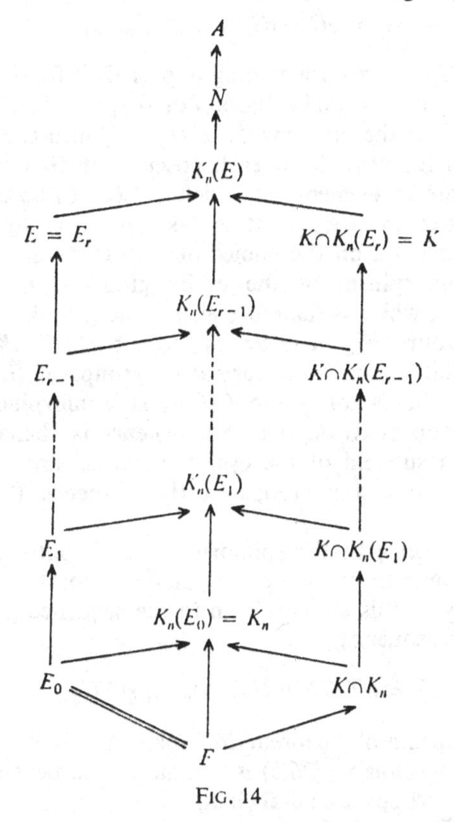

FIG. 14

which leaves K fixed is a normal subgroup of $\bar{G}$ and the factor group $\Gamma = \bar{G}/G$ is isomorphic to the Galois group of K over F. (All this follows from Theorem 16.3.) Since G is a normal subgroup of $\bar{G}$, the subgroup of $\bar{G}$ generated by G and the subgroup G_i is the product GG_i ($i = 0, \ldots, r$). Hence, according to Theorem 18.2, the sequence of subgroups of $\bar{G}$ corresponding to the sequence (26.3) of subfields of K is

$$G = GG_r, GG_{r-1}, \ldots, GG_0, \bar{G}. \tag{26.4}$$

Since G_{i+1} is a normal subgroup of G_i it follows easily that GG_{i+1} is a normal subgroup of GG_i ($i = 0, \ldots, r-1$). Now consider the mapping ϕ_i of G_i/G_{i+1} into GG_i/GG_{i+1} defined as follows: from each coset C of G_i relative to G_{i+1} choose an element τ and define $\phi_i(C)$ to be the coset of τ relative to GG_{i+1}. It is easy to verify that $\phi_i(C)$ does not depend on the choice of τ from C, and that ϕ_i is an epimorphism of the cyclic group G_i/G_{i+1} onto GG_i/GG_{i+1}, which is therefore also cyclic ($i = 0, \ldots, r-1$). The subgroup GG_0 of $\bar{G}$ leaves the subfield $K \cap K_n$ fixed; since G and G_0 are both normal subgroups of $\bar{G}$, so also is GG_0. The factor group $\bar{G}/GG_0$ is isomorphic to the Galois group of $K \cap K_n$ over F and hence is abelian, since $K \cap K_n$ is a subfield of the cyclotomic extension K_n of F. Thus all the factor groups in the sequence (26.4) are abelian.

Finally, let η be the epimorphism of $\bar{G}$ onto $\bar{G}/G = \Gamma$ which assigns to every element of $\bar{G}$ its coset relative to G. Applying this epimorphism to the sequence (26.4), we obtain the sequence

$$\{\varepsilon\} = \eta(GG_r), \eta(GG_{r-1}), \ldots, \eta(GG_0), \Gamma. \tag{26.5}$$

The arguments of Theorem 26.3, part (2), show that each of the subgroups in (26.5) is normal in the next and that the factor groups are all abelian.

Thus Γ is a solvable group, as we asserted.

§ 27. **Generic polynomials.** In this section, as in the preceding, we consider only fields of characteristic zero.

Let F be a subfield of a field E. A set $\{x_1, \ldots, x_n\}$ consisting of n elements of E is said to be **algebraically independent** over F if there are no polynomial relations with coefficients in F connecting the elements $x_1, \ldots, x_n$. More precisely, if we write $\mathbf{x}$ for the ordered n-tuple $(x_1, \ldots, x_n)$ we may say that the set $\{x_1, \ldots, x_n\}$ is algebraically independent over F if the only polynomial f in $P_n(F)$ such that $\sigma_{\mathbf{x}}(f) = 0$ is the zero polynomial. Thus, for example, if E is any field including the nth order polynomial ring $P_n(F) = F[X_1, \ldots, X_n]$, the set $\{X_1, \ldots, X_n\}$ is algebraically independent over F.

If now $\{x_1, \ldots, x_n\}$ is a subset of a field E algebraically independent over the subfield F of E, the polynomial

$$g_n = X^n - x_1X^{n-1} + x_2X^{n-2} - \ldots + (-1)^n x_n$$

in $P(F(\mathbf{x}))$ is called a **generic polynomial** of degree n over F. So a generic polynomial over F is one which has no polynomial relations with coefficients in F connecting its coefficients; thus we may think of it roughly as a " prototype " for the polynomials with coefficients in F, which can all be obtained from it on replacing the coefficients $x_1, \ldots, x_n$ by elements of F.

One of the problems which interested the classical algebraists was that of obtaining " formulæ " for the roots of generic polynomials. By a " formula " they meant what we should describe in modern terminology as an element of an extension by radicals of the field $F(\mathbf{x})$—that is to say, an element of some extension field of $F(\mathbf{x})$ which can be derived from the elements of F and the coefficients $x_1, \ldots, x_n$ by means of the field operations and the extraction of roots. As every schoolboy knows, the classical algebraists solved their problem in the case of generic polynomials of degree 2: the roots of

$$g_2 = X^2 - x_1X + x_2$$

are

$$\tfrac{1}{2}(x_1 + (x_1^2 - 4x_2)^{\frac{1}{2}}) \text{ and } \tfrac{1}{2}(x_1 - (x_1^2 - 4x_2)^{\frac{1}{2}}).$$

They solved it also in the case of generic polynomials of degrees 3 and 4. But they failed to find formulæ for the roots of generic polynomials of degree 5 or higher.

We shall now show that this failure was inevitable; we shall prove, in fact, that generic polynomials of degree greater than 4 are not solvable by radicals. According to Theorem 26.5, this result will follow if we can establish that for $n>4$ the Galois group of any splitting field of a generic polynomial g_n is not solvable. So we proceed first to determine the Galois group of a splitting field of a generic polynomial.

THEOREM 27.1. *Let* $g_n = X^n - x_1 X^{n-1} + \ldots + (-1)^n x_n$ *be a generic polynomial of degree n over a field F of characteristic zero. Then the Galois group of any splitting field of* g_n *over* $F(x_1, \ldots, x_n) = F(\mathbf{x})$ *is isomorphic to the symmetric group on n digits.*

Proof. Let K be a splitting field for g_n over $F(\mathbf{x})$; we may assume that K includes $F(x)$. Let $\alpha_1, \ldots, \alpha_n$ be the roots of g_n in K; then g_n splits completely in $P(K)$ in the form

$$\begin{aligned} g_n &= X^n - x_1 X^{n-1} + \ldots + (-1)^n x_n \\ &= (X-\alpha_1)(X-\alpha_2)\ldots(X-\alpha_n), \end{aligned}$$

from which we deduce that

$$\begin{aligned} x_1 &= \alpha_1 + \alpha_2 + \ldots + \alpha_n \\ x_2 &= \alpha_1\alpha_2 + \alpha_1\alpha_3 + \ldots + \alpha_{n-1}\alpha_n \\ &\ldots \\ x_n &= \alpha_1\alpha_2\ldots\alpha_n. \end{aligned}$$

We now define elements $e_1, \ldots, e_n$ of $F(\mathbf{x})$ by setting

$$\begin{aligned} e_1 &= x_1 + x_2 + \ldots + x_n \\ e_2 &= x_1x_2 + x_1x_3 + \ldots + x_{n-1}x_n \\ &\ldots \\ e_n &= x_1x_2\ldots x_n, \end{aligned}$$

and we write $e = (e_1, ..., e_n)$. (In the terminology of classical works on algebra, the elements x_i are the elementary symmetric functions of $\alpha_1, ..., \alpha_n$ and the elements e_i are the elementary symmetric functions of $x_1, ..., x_n$.) We claim that the set $\{e_1, ..., e_n\}$ is algebraically independent over F. So suppose there is a polynomial f in $P_n(F)$ such that $\sigma_e(f) = 0$. If $f = \Sigma a_{i_1...i_n} X_1^{i_1}...X_n^{i_n}$, with $a_{i_1...i_n}$ in F, let f^* be the polynomial

$$\Sigma a_{i_1...i_n}(\Sigma X_\nu)^{i_1}(\Sigma X_\nu X_\mu)^{i_2}...(X_1...X_n)^{i_n};$$

then clearly $\sigma_x(f^*) = \sigma_e(f) = 0$. Now, since the set $\{x_1, ..., x_n\}$ is algebraically independent over F, each element w of the polynomial ring $F[x] = F[x_1, ..., x_n]$ can be expressed in exactly one way in the form

$$w = \Sigma b_{i_1...i_n} x_1^{i_1}...x_n^{i_n};$$

so we may define a mapping ϕ of $F[x]$ into K by setting

$$\phi(w) = \phi(\Sigma b_{i_1...i_n} x_1^{i_1}...x_n^{i_n}) = \Sigma b_{i_1...i_n} \alpha_1^{i_1}...\alpha_n^{i_n}$$

for every element w of $F[x]$, and it is easy to verify that ϕ is a homomorphism. Hence $\phi(\sigma_x(f^*)) = 0$; but $\phi(\sigma_x(f^*)) = \sigma_\alpha(f^*) = \sigma_x(f)$. Thus $\sigma_x(f) = 0$ and so f is the zero polynomial; thus $\{e_1, ..., e_n\}$ is algebraically independent over F.

According to Theorem 8.1, every element of $F(x)$ can be expressed in the form $\sigma_x(p)/\sigma_x(q)$ where p and q are polynomials in $P_n(F)$ and q is not the zero polynomial. If q is not the zero polynomial, it follows from the algebraic independence of $\{e_1, ..., e_n\}$ over F that $\sigma_e(q)$ is non-zero; so we may form the element $\sigma_e(p)/\sigma_e(q)$ of $F(e_1, ..., e_n) = F(e)$. Further, it is easy to verify that if $\sigma_x(p)/\sigma_x(q) = \sigma_x(p')/\sigma_x(q')$, then $\sigma_e(p)/\sigma_e(q) = \sigma_e(p')/\sigma_e(q')$. Hence we may define a mapping τ of $F(x)$ into $F(e)$ by setting

$$\tau(\sigma_x(p)/\sigma_x(q)) = \sigma_e(p)/\sigma_e(q)$$

for every element $\sigma_x(p)/\sigma_x(q)$ of $F(x)$. It is easily seen that τ is an isomorphism of $F(x)$ onto $F(e)$.

Let τ_P be the canonical extension of τ to the polynomial ring $P(F(x))$; then we have

$$\tau_P(g_n) = X^n - e_1 X^{n-1} + e_2 X^{n-2} - \ldots + (-1)^n e_n.$$

The polynomial $\tau_P(g_n)$ splits completely in $P(F(x))$—namely,

$$\tau_P(g_n) = (X - x_1)(X - x_2)\ldots(X - x_n).$$

It follows that $F(x)$ is a splitting field for $\tau_P(g_n)$ over $F(e)$. According to Theorem 11.2, there exists an extension of τ which maps K isomorphically onto $F(x)$. Hence, by Theorem 16.4, the Galois group of K over $F(x)$ is isomorphic to the Galois group of $F(x)$ over $F(e)$.

This latter Galois group, however, is clearly isomorphic to the symmetric group on n digits. First of all, every element of the Galois group must effect a permutation of the roots $x_1, \ldots, x_n$ of $\tau_P(g_n)$, and no two distinct automorphisms effect the same permutation; but, conversely, since the set $\{x_1, \ldots, x_n\}$ is algebraically independent over F, every permutation of $x_1, \ldots, x_n$ induces an automorphism of $F(x)$ which clearly leaves $F(e)$ fixed. This one-to-one correspondence between permutations and automorphisms is easily seen to be an isomorphism of the symmetric group S_n onto the Galois group.

This completes the proof.

The result which we announced just before stating the theorem now follows at once. For, if we recall the remark in § 26 that S_n is not solvable when $n>4$, it follows that for any splitting field of g_n $(n>4)$, the Galois group is not solvable. Appealing to Theorem 26.5 we obtain our final theorem.

THEOREM 27.2. *Generic polynomials of degree greater than 4 are not solvable by radicals.*

Finis coronat opus.

Examples IV

1. Let F be a field of characteristic p with p^n elements. Prove that the additive group F^+ of F is isomorphic to the direct product of n cyclic groups of order p.

2. Let F be a finite field with q elements, K an extension of degree n over F. Show that there are $(q^n-1)/(q-1)$ elements x of K such that $N_{K/F}(x) = e$ and deduce that the norm mapping $N_{K/F}$ maps K onto F.

3. Let F be a finite field with q elements. If m is a positive integer relatively prime to q and K_m is a splitting field of $X^m - e$ over F, show that the degree $(K_m : F)$ is the least positive integer n such that $q^n - 1$ is divisible by m.

4. Let F be a finite field with q elements, m a positive integer relatively prime to q and Φ_m the mth cyclotomic polynomial with coefficients in the prime field of F. Show that Φ_m is irreducible in $P(F)$ if and only if the order of the residue class of q modulo m in the group $\mathbf{R}_m$ is $\phi(m)$.

5. Let K be a normal separable extension of finite degree over a subfield F. If f is a mapping of the Galois group G into the multiplicative group K^* of non-zero elements of K such that

$$f(\sigma) \,.\, \sigma(f(\tau)) = f(\sigma\tau)$$

for all pairs of elements σ, τ of G, prove that there exists an element a of K^* such that $f(\sigma) = a/\sigma(a)$ for all elements σ of G. [By Theorem 14.1 the mapping $\phi = \sum_{\rho} f(\rho) \,.\, \rho$ is not the zero mapping of K into K; take $a = \phi(b)$ where b is any element of K such that $\phi(b)$ is non-zero.] Deduce Hilbert's Theorem 90 in the case where K is a cyclic extension of F.

6. Let F be a field of characteristic zero such that $X^n - e$ splits completely in $P(F)$. Let $K = F(\alpha)$ be a cyclic

extension of degree n over F, where α is a root of the irreducible polynomial $X^n - a$ in $P(F)$. Show that an element β of K is a root of an irreducible polynomial of the form $X^n - b$ in $P(F)$ if and only if $\beta = c\alpha^r$ where r is an integer and c is an element of F.

7. Let F be a field of characteristic p, $K = F(\alpha)$ a cyclic extension of degree p over F, where α is a root of the irreducible polynomial $X^p - X - a$ in $P(F)$. Show that an element β of K is a root of a polynomial $X^p - X - b$ in $P(F)$ if and only if $\beta = k\alpha + \lambda$ where k is an integer and λ is an element of F.

8. Let $K = \mathbf{Q}(\eta)$ be the subfield of $\mathbf{C}$ generated over $\mathbf{Q}$ by adjoining the complex number $\eta = \cos \frac{2}{5}\pi + i \sin \frac{2}{5}\pi$. Show that K is a normal extension of degree 4 over $\mathbf{Q}$ with cyclic Galois group generated by the automorphism which maps η onto η^2. Prove that K has a unique subfield E of degree 2 over $\mathbf{Q}$ and that $\xi = \eta + \eta^4$, $\xi' = \eta^2 + \eta^3$ form a normal basis for E over $\mathbf{Q}$. Find the minimum polynomials of ξ, η over $\mathbf{Q}$, E respectively and hence express η in terms of radicals.

READING LIST

ALBERT, A. A., *Modern Higher Algebra*, Cambridge University Press, 1938.

ARTIN, E., *Galois Theory*, University of Notre Dame, 1948.

BOURBAKI, N., *Algèbre* (Ch. V, *Corps Commutatifs*), Hermann, 1950.

GAAL, L., *Classical Galois Theory, with Examples*, Chelsea, 1973.

HADLOCK, C. R., *Field Theory and its Classical Problems*, Mathematical Association of America, 1978.

JACOBSON, N., *Lectures in Abstract Algebra*, vol. 3 – *Theory of Fields and Galois Theory*, Van Nostrand, 1964.

KAPLANSKY, I., *Fields and Rings*, University of Chicago Press, 1969.

MCCARTHY, P. J., *Algebraic Extensions of Fields*, Blaisdell, 1966.

NAGATA, M., *Field Theory*, Dekker, 1977.

POSTNIKOV, M. M., *Foundations of Galois Theory*, Pergamon Press, 1962.

STEINITZ, E., *Algebraische Theorie der Körper*, Chelsea, 1950.

STEWART, I., *Galois Theory*, Chapman and Hall, 1973.

TSCHEBOTARÖW, N. and SCHWERDTFEGER, H., *Grundzüge der Galois'schen Theorie*, Noordhoff, 1950.

VAN DER WAERDEN, B. L., *Modern Algebra*, vol. 1, Ungar, 1949.

WINTER, D. J., *The Structure of Fields*, Springer, 1974.

ZARISKI, O. and SAMUEL, P., *Commutative Algebra*, vol. 1, Van Nostrand, 1958.

INDEX OF NOTATIONS

INDEX